HN 25 .Q35

Quality-of-life community indicators for parks

 **W9-AHH-402**

DATE DUE

OCT 2 1	2011		

DEMCO 38-296

Quality-of-Life Community Indicators for Parks, Recreation and Tourism Management

Social Indicators Research Series

Volume 43

General Editor:

ALEX C. MICHALOS
University of Northern British Columbia,
Prince George, Canada

Editors:

ED DIENER
University of Illinois, Champaign, U.S.A.

WOLFGANG GLATZER
J.W. Goethe University, Frankfurt am Main, Germany

TORBJORN MOUM
University of Oslo, Norway

MIRJAM A.G. SPRANGERS
University of Amsterdam, The Netherlands

JOACHIM VOGEL
Central Bureau of Statistics, Stockholm, Sweden

RUUT VEENHOVEN
Erasmus University, Rotterdam, The Netherlands

This new series aims to provide a public forum for single treatises and collections of papers on social indicators research that are too long to be published in our journal Social Indicators Research. Like the journal, the book series deals with statistical assessments of the quality of life from a broad perspective. It welcomes the research on a wide variety of substantive areas, including health, crime, housing, education, family life, leisure activities, transportation, mobility, economics, work, religion and environmental issues. These areas of research will focus on the impact of key issues such as health on the overall quality of life and vice versa. An international review board, consisting of Ruut Veenhoven, Joachim Vogel, Ed Diener, Torbjorn Moum, Mirjam A.G. Sprangers and Wolfgang Glatzer, will ensure the high quality of the series as a whole.

For further volumes:
http://www.springer.com/series/6548

Megha Budruk · Rhonda Phillips
Editors

Quality-of-Life Community Indicators for Parks, Recreation and Tourism Management

 Springer

Editors
Dr. Megha Budruk
Arizona State University
School of Community
Resources & Development
Mail Code 4020
N. Central Ave. 411
85004-0690 Phoenix Arizona
Suite 550
USA
Megha.Budruk@asu.edu

Dr. Rhonda Phillips
Arizona State University
School of Community
Resources & Development
Mail Code 4020
N. Central Ave. 411
85004-0690 Phoenix Arizona
Suite 550
USA
rhonda.phillips@asu.edu

ISSN 1387-6570
ISBN 978-90-481-9860-3   e-ISBN 978-90-481-9861-0
DOI 10.1007/978-90-481-9861-0
Springer Dordrecht Heidelberg London New York

Printed on acid-free paper

Springer is part of Springer Science+Business Media (www.springer.com)

"When we all do better, we all do better."

– Paul Wellstone

For my parents Shripal and Aruna

- Megha Budruk

To my colleagues, for their support and inspiration

- Rhonda Phillips

Contents

Contributors

Kathleen L. Andereck Arizona State University, Phoenix, AZ, USA,
Kathleen.Andereck@asu.edu

Megha Budruk Arizona State University, Phoenix, AZ, USA,
Megha.Budruk@asu.edu

HwanSuk Chris Choi University of Guelph, Guelph, ON, Canada,
Hwchoi@uoguelph.ca

Sam Cole University of Buffalo, Buffalo, NY, USA, samcole@buffalo.edu

Bill Field International Languages Department, University of Applied Science
Landshut, Am Lurzenhof 1, 84036 Landshut, Germany, bill.field@web.de

Sheila Flanagan Dublin Institute of Technology, Dublin, Ireland,
Sheila.flanagan@dit.ie

Kevin Griffin Dublin Institute of Technology, Dublin, Ireland,
Kevin.griffin@dit.ie

Robert E. Manning University of Vermont, Burlington, VT, USA,
Robert.Manning@uvm.edu

Roland Mányai Department of Tourism, Budapest, Hungary,
Manyai.roland@otm.gov.hu

Maeve Morrissey Dublin Institute of Technology, Dublin, Ireland,
Maeve.morrissey@dit.ie

Donna Myers Arizona State University, Phoenix, AZ, USA,
Donna.M.Myers@asu.edu

Jeremy Németh University of Colorado, Denver, CO, USA,
jeremy.nemeth@colorado.edu

Gyan Nyaupane Arizona State University, Phoenix, AZ, USA, Gyan@asu.edu

Rhonda Phillips Arizona State University, Phoenix, AZ, USA,
rhonda.phillips@asu.edu

László Puczkó Corvinus University, Budapest, Hungary, Lpuczko@xellum.hu

Victoria Razak University of Buffalo, Buffalo, NY, USA, vrazak@buffalo.edu

Ariel Rodríguez Arizona State University, Phoenix, AZ, USA,
Ariel.Rodriguez@asu.edu

Stephan Schmidt Cornell University, Ithaca, NY, USA, Sjs96@cornell.edu

Melanie Smith Corvinus University, Budapest, Hungary,
Melanie.smith@uni-corvinus.hu

Ercan Sirakaya Turk University of South Carolina, Columbia, SC, USA,
Ercan@mailbox.sc.edu

About the Contributors

Kathleen L. Andereck is the director and a professor in the School of Community Resources and Development at Arizona State University, USA. Her primary focus is the study of the tourism experience from the perspective of both tourists and community residents. She has conducted recreation- and tourism-related research projects for many entities including the USDA Forest Service, Bureau of Land Management, National Park Service, Arizona Office of Tourism, Arizona Department of Transportation, Arizona Department of Commerce, and several Arizona communities. Dr. Andereck is an associate editor for four academic journals and is the author of many refereed journal articles and numerous conference papers and professional reports. She can be contacted at Kathleen.Andereck@asu.edu.

Megha Budruk is an associate professor in the Parks and Recreation Management Program, School of Community Resources and Development, Arizona State University, USA. Her research interests include human–natural environment relationships. Specifically, Dr. Budruk's research focuses around place attachment or the special bonds between humans and natural places, crowding and indicator-based social carrying capacity, as well as beneficial outcomes of natural resource-based recreation. She has published in several peer-reviewed journals including *Society and Natural Resources, Journal of Leisure Research, Journal of Park and Recreation Administration*, and *Environmental Management*. She can be contacted at Megha.Budruk@asu.edu.

HwanSuk Chris Choi is an associate professor at the School of Hospitality & Tourism Management, University of Guelph, Canada. His broad research interests include areas of visitor satisfaction, destination identity, post-experience emotion, sustainable tourism management, and research methodology in hospitality and tourism. He can be contacted at Hwchoi@uoguelph.ca.

Sam Cole is a professor in the Department of Urban and Regional Planning at the University of Buffalo, USA, and a research fellow of the Center for Urban Studies, teaching and researching economic development, futures studies, and tourism. Originally trained as a theoretical physicist, Dr. Cole has published in several fields and worked for a variety of national and international organizations. He won several national planning awards including for "Dare to Dream: Bringing Futures Studies into Planning," the JAPA Best Paper 2000, with Dowell Myers and others, and, with Henry Taylor, the 2001 Fannie Mae Foundation Award, Best Action Research Paper. In 2006, his students won the APA Carole R. Bloom Award for best tourism studio for their socioeconomic analysis of recreational second homes. He can be contacted at samcole@buffalo.edu.

Bill Field has an extensive background as an instructor and business trainer in fields as diverse as natural resource management, marketing, finance, technical studies, and intercultural relations. Since his arrival in Germany, he has been working as a university instructor, as well as in training or consulting roles for numerous international clients. He has been responsible for developing both E-learning and classroom-based instructional programs for academic as well as corporate clients. Bill pursued multiple degrees in communications, community economic development, and natural resources and environmental studies. This background has assisted him in working as a consultant in tourism, value-added small enterprise development, financial services, and microfinance. Bill has worked and lived in three continents where he has continued to pursue his interest in sustainable economic development and training. He can be contacted at bill.field@web.de.

Sheila Flanagan, graduate of University College Dublin, is currently head of School of Hospitality Management and Tourism and senior lecturer in Tourism Policy and Planning in the DIT. She was responsible for the development of the first masters-level program in tourism management on this island, through the design of the MBS in tourism for the Smurfit Graduate Business School in UCD. Dr. Flanagan has considerable industry experience in tourism planning in both public and private bodies, national and international, including the European Union, Irish Government Departments, and the Irish Tourist Board and management consultants. She is incoming president of Travel & Tourism Research Association (TTRA) International and past president of TTRA Europe and is currently the project director on a major EPA-funded research project into the development and implementation of sustainable tourism indicators. In brief her portfolio includes the following:

Twenty years academic experience in tourism planning and destination management, substantial peer-reviewed papers in these areas, expertise in research supervision – at undergraduate, postgraduate, and Ph.D. level, tourism planning consultant for national and international public and private bodies, development of tourism strategy plans at community, county, and regional levels, pioneer of the usage of bottom up and integrated rural tourism planning methodologies in Eastern Europe, technical advisor to European City Tourism (ECT), central involvement in the implementation of the Dublin component of a Europe wide city survey, and leader in the recent branding initiative – MagicTouch Partners. Her school is the first to develop innovative partnerships with industry. She can be contacted at Sheila.flanagan@dit.ie.

Kevin Griffin, head of the Department of Tourism in the Dublin Institute of Technology (DIT), is an experienced educator with research interests in academic, local studies, and commercial spheres. With a twin background in geography (tourism/historical) and education, his research has always been both academic and practical field based. With research interests ranging from sustainability to heritage management and local history to religious tourism, he is interested predominantly in research that has meaning and purpose, particularly to those on the ground, especially at a local community level. Kevin was a strand manager on a major EPA-funded project which identified a model of sustainable tourism indicators and is a research leader on a follow-on EPA-funded project. Kevin is a founder member and on the committee of the DIT Sustainability Research Group – an interdisciplinary cross-faculty research initiative of DIT, which seeks to bring together teaching and research initiatives of a sustainable nature. He has been actively involved in embedding sustainability principles in a number of undergraduate- and postgraduate-level modules which he teaches. He can be contacted at Kevin.griffin@dit.ie.

Robert E. Manning is a professor in the Rubenstein School of Environment and Natural Resources at the University of Vermont, USA, where he teaches courses in park and wilderness management and environmental history and philosophy. He conducts a program of research for the U.S. National Park Service. Dr. Manning is the author of numerous refereed journal articles and conference papers as well as several books including *Parks and Carrying Capacity: Commons without Tragedy* and *Studies in Outdoor Recreation*. He can be contacted at Robert.Manning@uvm.edu.

Roland Mányai is the head of department, Hungarian Ministry of Local Government, Tourism Unit. He can be contacted at Manyai.roland@otm.gov.hu.

Maeve Morrissey is a graduate of Dublin Institute of Technology (M.Sc. Sustainable Development) and of University of Dublin, Trinity College (B.A. (Hons) Microbiology). Her areas of particular interest include sustainable tourism, sustainable development, strategic environmental assessment (SEA), environmental impact assessment (EIA), and environmental management. She is a graduate member of IEMA. Maeve has previously worked for an environmental consultancy, where she was involved in sustainability projects for tropical rainforests based on sustainable tourism and "Reducing Emissions from Deforestation and Forest Degradation" (REDD) and on the EIA's of large infrastructural projects. She has also worked in the clinical trial industry where she was responsible for project management of the data management aspect of drug trials. Maeve's role in the Dublin Institute of Technology is as project coordinator for the DIT-ACHIEV Project for achieving sustainable tourism management, which is jointly funded by Fáilte Ireland and the Irish Environmental Protection Agency. She can be contacted at Maeve.morrissey@dit.ie.

Donna Myers is a graduate alumnus, School of Community Resources and Development, Arizona State University, USA. Her research focused on community development and tourism with a particular interest in the role of community indicator systems to help achieve citizen participation and sustainable tourism outcomes. She holds an M.S. in Recreation and Tourism Studies. She can be contacted at Donna.M.Myers@asu.edu.

Jeremy Németh is an assistant professor in the Department of Planning and Design at the University of Colorado Denver, USA, where he is also the director of the urban design program. His research interests include the privatization of public space, perceptions of the built environment, and the legal geography of protest. He can be contacted at jeremy.nemeth@colorado.edu.

Gyan Nyaupane is the graduate program director and an assistant professor at Arizona State University, USA. Dr. Nyaupane has research interests in understanding the relationship between tourists and natural and cultural resources, and tourism's contribution toward environmental conservation and poverty alleviation. He has done extensive research on topics related to ecotourism, nature-based and heritage tourism, and user fees and has published his findings in major tourism and recreation journals including *Annals of Tourism Research, Journal of Travel Research, Leisure Sciences, Tourism Management,* and *Journal of Sustainable Tourism.* He has recently co-edited a book on heritage and tourism in the developing world. He can be contacted at Gyan@asu.edu.

Rhonda Phillips is a professor in the School of Community Resources and Development, Arizona State University, USA. Her research interests include community and economic development, community indicator systems and quality of life, community well-being, sustainability, cultural/arts, and tourism development approaches. Prior books include *Introduction to Community Development* and *Community Development Indicators Measuring Systems.* Along with Joe Sirgy and Don Rathz, she is the co-editor of the *Community Quality-of-Life Indicators: Best Cases* series published through ISQOLS and Springer. She can be contacted at rhonda.phillips@asu.edu.

László Puczkó is the head of tourism at Xellum Ltd and professor at Budapest Corvinus University. Graduated in Business Administration at Budapest University of Economic Sciences (1993). He holds an M.A. in Art & Design Management (Hungarian Academy of Arts and Crafts) and a Ph.D. (Budapest University of Economics and Public Administration) and is a certified management consultant. Dr. Puczko is a board member of TTRA European Chapter. His main areas of expertise are tourism-specific quality of life, visitor management, and health and heritage tourism. He is a co-author of books on the impacts of tourism, visitor management, and tourism management in historic cities, as well as on health and wellness tourism. He can be contacted at Lpuczko@xellum.hu.

Victoria Razak is a cultural and visual anthropologist. Her teaching at the University at Buffalo, USA, in the Department of American Studies and Department of Urban Planning includes qualitative research methods, immigrant settlement patterns, and world civilizations. Her principal research interests include Caribbean identity politics and migrant communities; indigenous music and carnival in Aruba; and cultural heritage tourism. Recent publications include "Carnival in Aruba: A Feast of Yourself" (in Green, Indiana University Press, 2007) and "From Culture Areas to Ethnoscapes: An Application to Tourism Development" (Journal of Regional Studies, 2008). Dr. Razak is also a research associate of the University of Aruba and a fellow of the University at Buffalo Center for Urban Studies. She can be contacted at vrazak@buffalo.edu.

Ariel Rodríguez is an assistant professor in the Parks and Recreation Management Program, School of Community Resources and Development, Arizona State University, USA. His research interests include leisure, quality of life, life satisfaction, physical activity, obesity, physical and mental health, and youth populations. He can be contacted at Ariel.Rodriguez@asu.edu.

Stephan Schmidt is an assistant professor in the Department of City and Regional Planning at Cornell University, USA. His research interests include the formation of local environmental policy, the social and political context of public spaces, and international comparative planning practices. He can be contacted at Sjs96@cornell.edu.

Melanie Smith is a senior lecturer and researcher in tourism in the Faculty of Public Administration at Corvinus University, Budapest. Before this, she was the director of B.A. Tourism and M.A. Cultural Tourism Programmes at the University of Greenwich in London. She is also the Chair of ATLAS (Association for Tourism and Leisure Education). She is the author of several books and journal articles with research interests in cultural tourism, urban regeneration, health and wellness tourism, and quality of life. She can be Melanie.smith@uni-corvinus.hu.

Ercan Sirakaya Turk is a sloan research professor of tourism at the School of Hotel, Restaurant and Tourism Management, University of South Carolina, USA. His research interests include tourism policy and development through sustainable tourism, international tourism and marketing, and destination marketing and management. He can be contacted at Ercan@mailbox.sc.edu.

Chapter 1
Introduction

Rhonda Phillips and Megha Budruk

Background

Within leisure research, approaches to measuring quality of life (QOL) have often focused on place-centered indicators such as the frequency of leisure facility usage. Lloyd and Auld (2002) propose that person-centered criteria such as satisfaction with leisure experiences or attributes of leisure need to be included as well in order to offer a more comprehensive view. An underlying assumption of place-centered indicators seems to be for "policy outcomes that increasing the number of facilities and services will automatically enhance people's QOL" (p. 43). In contrast, Lloyd and Auld's research found that person-centered measures are the best predictors of quality of life. This supports other findings that leisure participation improves personal quality of life in a variety of ways (Coleman & Iso-Ahola, 1993). This includes "holidaytaking" or vacationing and its uplifting effects (Gilbert & Abdullah, 2004) or increases in recreational/physical activity and improved well-being (Wendel-Vos, Schuit, Tijhuis, & Kromhout, 2004). These are just a few of numerous studies that show these positive benefits. So, we know that leisure is a vital component of quality of life.

Leisure and quality of life is a complex and fascinating domain of study and one that deserves much attention. An interesting point to consider is how the parks, recreation, and tourism field can inform ongoing research and inquiry into a person-centered perspective of quality of life and community development. As Crouch and Ritchie (1999, p. 189) point out, much of "tourism development entails community development. . . . If handled appropriately, tourism can become an important engine for achieving broader social goals."

The latter part of the quote, "achieving broader social goals," represents a shift in perspectives of what tourism and parks and recreational aspects of communities really mean and how these are valued. This desire for achieving broader goals is

R. Phillips (✉)
Arizona State University, Phoenix, AZ, USA
e-mail: rhonda.phillips@asu.edu

M. Budruk, R. Phillips (eds.), *Quality-of-Life Community Indicators for Parks, Recreation and Tourism Management*, Social Indicators Research Series 43, DOI 10.1007/978-90-481-9861-0_1, © Springer Science+Business Media B.V. 2011

evident in more recent studies where indexes are constructed to gauge impacts that cover a wide range of community well-being considerations. For example, Urtasun and Gutiérrez's (2006) study of numerous Spanish provinces included such indicator indexes as health and health services, cultural and leisure opportunities, coexistence and participation, and citizen security in addition to more typical social, economic, and environmental indicators. Similarly, Roberts and Tribe's (2008) work in sustainable tourism indicators point out the importance of constructing indicators and indexes to address residents' and tourists' concerns and areas of conflict. These areas include resident access to recreational and tourist areas, promotion and protection of indigenous cultures, and ownership patterns of venues.

Community indicators represent one such way to bridge these areas of quality of life and broader social goals, especially when used to address areas of concern that are important to residents, visitors, and organizations – the stakeholders of our communities. Indicators as well as standards have long been used for management decision making in the leisure sciences. We see particularly strong evidence of their use within natural resource management, where environmental and social indicators are relied upon in maintaining resource and visitor experience integrity.

Since the passage of the USA's National Environmental Policy Act of 1969, community participation in governmental decision-making processes has received increased awareness and legal mandate. Community indicators provide one avenue for community voices to be integrated in the planning and management of common resources. Within the tourism management domain, more community indicator frameworks are being developed and used from the local and regional to national and multinational levels. Our area of interest is at this intersection – where quality-of-life domains and community indicators can be used for management of parks, recreation, and tourism concerns. It is not just about indicators or measurement of quality of life; however, it is about how these are integrated into "systems" that influence management processes and outcomes for the overall well-being of communities. Simply having indicators is not enough, it is their use and integration into larger decision-making and policy processes that subsequently influences personal and overall community quality of life.

What Is So Special About Community Indicators?

A bit of context for community indicators is merited. What makes community indicators any different from other measures? Their specialness lies in the ability to help build an integrative approach – considering impacts in not only economic terms, but also the social and environmental dimensions. It is the ability to build a system or framework on valid indicators that conveys its real usefulness. These systems can then be used to aid decision making and set priorities within organizations and communities. Further, a community indicator's system reflects collective values and this is a powerful feature. If the process of identifying and implementing indicators is open and inclusive, i.e., if citizen and stakeholder participation are embedded in the process, then the system will reflect collective values. Typically, decisions made on the basis of collective values receive more widespread support since collective values imply that goals or targets are more widely agreed upon. Finally, indicators' systems or frameworks represent a more comprehensive evaluation tool. Since these systems can be integrated into community development planning for an overall community or region's planning, it makes evaluation easier.

It is this ability of community indicators to be integrated as a system for gauging impacts across a full spectrum of considerations mentioned above that makes it beneficial to explore using them. Further, indicators incorporate both frameworks of performance and process outcomes, which serve to facilitate evaluation. When properly integrated into planning and development, they hold the potential to go beyond just issuing annual summary reports, to being utilized in the decision-making process. Widespread support or recognition of the role of parks, recreation, and tourism within a larger community and regional comprehensive planning context can be very beneficial for all.

A Closer Look at Indicators

Just what is a community "indicator"? Basically, community indicators are bits of information that when combined provide a picture of what is happening in a local, regional, or national system (Phillips & Pittman, 2009, p. 285). They provide insight into the direction of a community: improving or declining, forward or backward, increasing or decreasing. For example, the number of new tourism-related jobs created might inform a community about the economic impacts of tourism development in their community. Combining indicators provides a measuring system or framework as mentioned previously to provide clear and honest information about past trends, current realities, and future direction, in order to aid decision making. Community indicators can also be thought of as a "report card" of community well-being, or a bottom line, similar to a balance sheet snapshot of a company. It is important to note that these systems generate much data and it is the analysis of these data that can be used in the decision-making and policy/program improvement process. There are four common frameworks used for developing and implementing community indicator systems in the USA. Each has relevance for parks, recreation, and tourism management and often a combination of the approaches is used to fully address needs and desires:

1 quality of life,
2 performance evaluation,
3 healthy communities, and
4 sustainability.[1]

A brief look at each of these illustrates the types of applications for community indicators.

Quality of Life

Quality of life is reflective of the values that exist in a community. Indicators therefore could be used to promote a particular set of values by making clear that residents' qualify of life is of vital importance. The advantage of this type of system is that is if agreement can be reached, the system can be strong motivator for all types of community outcomes, not the least of which is evaluating progress toward common goals. The disadvantage is that measuring quality of life is a political process because what defines "good life" can vastly differ among individuals, groups, and institutions. See the subsequent discussion of quality of life for additional insights.

Performance Evaluation

This type is mostly managed by state or local governments and organizations as a way to indicate the outcomes achieved by their activities. It is very beneficial as an evaluative technique because it provides regular (usually annual or semi-annual) reports on progress and outcomes. It is typically part of the annual budgeting process so that adjustments can be made for priority areas. Performance evaluation is used often in parks, recreation, and tourism management, particularly when gauging impacts of programs.

Healthy Communities

This approach is gaining popularity as it attempts to cultivate a sense of shared responsibility for community health and well-being. It focuses on indicators that reflect health such as elder care, pre-natal and early childhood care, and youth – groups that often do not show on typical economic indicators concerned with working adults. It also prioritizes education and other human development facets of communities, including social concerns. These projects often redefine traditional measures of healthy conditions and typically build upon the World Health Organization's Healthy Cities program. This type of approach holds much relevance for parks and recreation management and can be seen in some applications that tie health policies with healthy living resources represented in parks and recreational resources.

Sustainability

"Community indicator systems" can provide the mechanism for monitoring progress toward balanced – "sustainable" – development because community indicators provide information for considering the impacts of development in not only economic terms but also social and environmental dimensions. The concept of sustainable development first emerged in the Bellagio Principles that include such characteristics as broad citizen participation, ongoing assessment, essential elements, and a guiding vision. Indicators are consistent with these principles. Indicators are just one means – but potentially a particularly effective means – of measuring progress toward sustainable development in a manner consistent with these principles. Indicators also offer the opportunity to go beyond standard indicators, such as gross domestic product, to fully assess well-being. The difficulty with the approach is being able to fully integrate indicator use into overall community planning so that sustainability can be a reality instead of rhetoric.

As mentioned, all of these approaches hold relevance for parks, recreation, and tourism management. Elements of each can be seen in many current approaches, and their applicability continues as the need to incorporate comprehensive community indicator systems expands.

Quality of Life Considerations

Because it is a special domain, we have to look further into quality of life concepts and issues. First, despite the different approaches used for identifying, designing, and implementing community indicator systems, all share the common value of improving the quality of living of people and places (Wong, 2006). Because QOL is "embedded" within indicators, it is important to consider its dimensions. There are numerous approaches to defining quality of life. It is not an easy task to identify and define it, much less measure it. It is an inherently political process because it involves competing ideologies that define what constitutes a "good life" in different ways (Phillips, 2003). Often, the utilitarianism model is used to valuate indicators (seen in rankings, for example, myriad annual "best places to live" studies). This model holds that individuals maximize their quality of life based on available resources and their individual desires and is driven by economic theory (Diener & Suh, 1997). However, there are limitations to its use as a guide for indicator development – for example there are concerns about its ability to be linked to actions and policies. Just because a community has a vibrant cultural or arts district does not necessarily mean citizens will be able to make use of it and thus positively impact their quality of life (Phillips, 2003).

Parkins et al. describe two other major approaches to defining and measuring quality of life, based on seminal work by Ed Diener and Eunkook Suh. The first is a normative approach based on commonly held beliefs or ideals within society and is most closely associated with the social indicators research tradition (Parkins, Stedman, & Varghese, 2001). Second, analysis is taken from the national or community level to the experience of the individual. It represents a union of social indicators research on objective human measures with a subjective approach of quality of life research (Parkins et al., 2001, p. 45). These approaches still lack the explicit reference to the issue of long-term balance; many suggest that sustainability is the bridging concept that will bring together human and ecological well-being together (Michalos, 1997; Parkins et al., 2001).

Another consideration is the need to look at "community" level quality of life, and this is more difficult. Is it simply a compilation of individuals' ratings? Therefore, we have to look at another distinguishing feature of quality of life: QOL measured through different units or levels of analysis and QOL measured through subjective versus objective indicators (Sirgy, Rahtz, Muris, & Underwood, 1998, p. 281). There is long-standing controversy over the latter, particularly with subjective indicators being used in several studies of community level indicators. Early on, Shin (1980 in Sirgy et al., 1998, p. 284) concluded that quality of life at the community level has two dimensions, the first is the level of citizen satisfaction related with various community resources and the second is the distribution of this satisfaction across the citizenry.

As seen in this brief discussion, measuring quality of life is not without its challenges, and issues of level and type of analysis are key. Despite the challenges, measuring quality of life is vital as it is especially reflective of the domains of parks, recreation, and tourism.

Bringing It Altogether: Quality of Life and Indicators

One way to bring together the diverse array of considerations in quality of life and community indicators is via strong planning. As mentioned, the ability to integrate indicator systems for parks, recreation, and tourism management into overall community, regional or national level planning is paramount to long-term success. And long-term success is all about sustainability and achieving balance. Since the early 1990s, the World Tourism Organization (WTO) has promoted the development and application of sustainability indicators for tourism destination communities, at the local to national levels. Their efforts have focused on the importance of incorporating indicators into planning processes and decision making. They explain as follows:

> Sustainable development of tourism requires a sound planning process, as well as continuous management of the key elements that support tourism and its destinations (e.g., maintenance of assets, involvement of the community, involvement of tourism in the planning process for the destination). Indicators are an intrinsic component of the planning process (World Tourism Organization, 2004, p. 14).

The WTO further describes how a community can respond with a planning process, whether or not they have current plans in place. It is an excellent illustration of how to integrate community indicators into planning. Here is an excerpt of this valuable information for incorporating indicators into tourism planning:

- Where no plan currently exists

    o The procedure by which indicators are developed is analogous to the first step in plan development. Both involve the identification of the key assets and key values associated with the destination. Both normally involve the assessment of the actual problems, current or potential impacts, or risks associated with development, as well as documentation of the major current or expected trend or events which may affect these.

o An indicators study can be the catalyst for development of a formal plan or planning process, beginning with identification of potential issues (pollution, loss of access, impacts of development in other sectors). Response will require some form of plan or management procedure.

o An indicators exercise can help identify key element that must be included in plans, such as the resource base for the industry or risks to assets or product

o Performance indicators can be defined relative to the specific goals and target of the plan; each specific development project can integrate performance indicators in order to measure the success of management actions in the implementation phase. This information will serve to decide whether corrective actions are needed and also can provide a tool for continuous monitoring.

o Indicators defined to analyze actual environmental and socio-economic conditions at the initial phase of the planning process can become performance indicators in the implementation phase. For example, indicators determining the actual state of seawater quality at beaches or actual levels of community income from tourism will serve later on to measure achievement relative to these goals.

• When a plan already exists

o An indicators study can assist in evaluation of current regional or tourism plans to determine where all of the key risks to sustainable development of tourism are covered

o The indicators identification exercise can be applied to already defined problems, issues, and objectives to improve the provision of accurate data and information where needed

o Where no monitoring system or performance measures are included in an existing plan, the indicators development process can assist in identifying and clarifying key area where performance measures are needed

o Indicators discussion can often stimulate greater precision in redefining goals and targets

(Source: World Tourism Organization, Madrid: WTO, 2004, pp. 14–15).

In summary, it is not an easy task to bring together all the dimensions of the elements addressed in this book: quality of life, sustainability, community indicators in the context of parks, recreation, and tourism management. For example, quality of life can be difficult to measure yet we must try for it as this aspect is highly valued and influences both the individual and the overall community or region. We know implicitly that it has major implications for parks, recreation, and tourism management. At the same time, community indicators can reflect quality of life and often explicitly represent it. Sustainability underlies all – having desirable quality of life as well as excellent management practices and outcomes of parks, recreation, and tourism results in communities, regions, and nations that are balanced for the long term. Indicator systems represent a tool to help capture the goals and targets for an area that are reflective of collective values and as such can be of great benefit.

Structure of This Book

Exploring all these concepts and bringing them together has been an enlightening process that we hope will spur innovative applications in the leisure sciences. Drawing from a variety of disciplines and perspectives such as parks, recreation, tourism, planning, and community development, we are pleased to present the contributions of 18 authors in 11 chapters. This first chapter provides an overview of quality of life and community indicators and sets the context for exploring them in the domains of the leisure sciences – parks and recreation and tourism. In Chapter 2, Robert Manning provides a review of indicators and standards in parks and outdoor recreation from the perspective of the environmental management field. His discussion of social norms and norm "intensity" has particular relevance for weighting the importance of potential indicators strengths.

Next we turn our attention in Chapter 3 to a philosophical exploration of the meanings of leisure and their relationship to quality-of-life satisfaction. Ariel Rodriguez presents a framework of leisure's three major meanings: leisure as free time, leisure as an activity, and leisure as an experience. Relating it to quality of life satisfaction underscores the importance of tackling the tough issue of measurement so that leisure can be included more broadly in indicator studies. As he points out, leisure has been a "slippery" concept throughout the quality of life and subjective well-being literature.

Drawing in a community planning perspective, Jeremy Nemeth and Stephan Schmidt present an indicator framework for urban public space management in Chapter 4. Creating a comprehensive index that uses 20 indicators in four broad categories, the authors demonstrate how it can be used to help manage publicly accessible spaces in our urban environments. An important feature of the index is that it can also serve to engage local communities and help facilitate public participation in the management process. In Chapter 5, Bill Field explores how to reconcile visitors' expectations with quality of life dimensions. His exploration is in the context of quality of life indicators for multi-use trail planning in an integrated resource management process. Using British Columbia's heritage trails as the test case, surveys based on expectancy theory are found to be a tool for measuring leisure quality of life values.

Beginning with Chapter 6, we turn our attention to the tourism domain. Kathleen Andereck and Gyan Nyaupane have developed a tourism quality-of-life instrument providing a comprehensive method of measuring residents' perceptions of tourism's impact on quality of life. The authors' identification of challenges and strengths in a variety of contexts is compelling and the subsequent analysis using exploratory factor analysis yields much insight. The instrument is a highly beneficial contribution to continuing efforts to integrate community indicators into tourism planning and allows for the inclusion of perceptions of life satisfaction which is the conceptual goal of quality of life studies. This chapter is followed by a classic reading with a reprint of HwanSuk Choi and Ercan Turk's "Sustainability Indicators for Managing Community-Based Tourism" in Chapter 7. This chapter helped foster the discussion of sustainability in the tourism context, by identifying 125 indicators via a panel of experts. The indicators are in the areas of political, social, ecological, economic, technological, and cultural dimensions. It brings to light the importance of indicators for community tourism development.

Chapter 8 by Sam Cole and Victoria Razak addresses the question of how far can tourism development proceed before the way of life enjoyed by residents is threatened by over development. Using the Caribbean Island of Aruba, the authors develop a framework to provide a structure for discussion of tourism-related sustainability issues. A very valuable aspect of this framework is its inclusion of carrying capacity indicators and serves as the baseline for discussing other economic, social, and demographic concerns.

A tourism-focused quality-of-life index for Budapest is the focus of Lazlo Puczko, Melanie Smith, and Roland Manyai's contribution in Chapter 9. The Budapest model introduces an approach to quality of life studies in the tourism context, bringing together five domains: attitudes toward traveling, motivations of the visitor, qualities of the trip, characteristics of the destination, and impacts of tourism. A comprehensive discussion of tourism and quality of life precedes the testing and discussion of the survey instruments for developing the model.

In Chapter 10, a case of a southwestern USA mining town, Mammoth, Arizona, Donna Myers, Megha Budruk, and Kathleen Andereck design a destination-level tourism indicator system focused on sustainability. The research was conducted from the perspectives of residents and business owners in the town. The indicators thus provide a voice for stakeholders in tourism development, reiterating the notion that indicators must be reflective of collective values of those they serve.

This volume concludes with Chapter 11, The Trials and Tribulations of Implementing Indicator Models for Sustainable Tourism Management: Observations from Ireland. Kevin Griffin, Maeve Morrissey, and Sheila Flanagan of the Dublin Institute of Technology outline the development of indicator model for sustainable management of tourism. The chapter includes a discussion on its design, testing, and implementation and how communities have been involved in these processes. The indicator model is quite comprehensive and represents an innovative application of community indicators and tourism planning and management, incorporating aspects of heritage, infrastructure, and enterprise.

A range of applications are evident in this collection of chapters from technically-led initiatives to identify identifying indicators (e.g., Chapters 6 and 7) to participatory/community-led initiatives (e.g., Chapters 8 and 10). These chapters illustrate how different situations warrant different approaches toward indicator applications. It is our sincere desire that you will find this volume useful to bring together quality-of-life considerations and community indicators in your own applications for inspired parks, recreation, and tourism management research and practice.

Note

1. A full description of these frameworks was first presented in *Community Indicators*, PAS Report Number 517, by Rhonda Phillips, published by the American Planning Association, 2003.

References

Coleman, D., & Iso-Ahola, S. E. (1993). *Journal of Leisure Research, 25*(2), 111–128.

Crouch, G. I., & And Ritchie, J. R. B. (1999). Tourism, competitiveness, and societal prosperity. *Journal of Business Research, 44*, 137–152.

Diener, E., & Suh, E. (1997). Measuring quality of life: Economic, social and subjective indicators. *Social Indicators Research, 40*(1), 189–216.

Gilbert, D., & Abdullah, J. (2004). Holidaytaking and the sense of well-being. *Annals of Tourism Research, 31*(1), 103.

Lloyd, K. M., & Auld, C. J. (2002). The role of leisure in determining quality of life: Issues of content and measurement. *Social Indicators Research, 57*, 43–71.

Michalos, A. (1997). Combining social, economic and environmental indicators for measuring sustainable human well-being. *Social Indicators Research, 40*, 221–258.

Parkins, J. R., Stedman, R. C., & Varghese, J. (2001). Moving towards local-level indicators of sustainability in forest-based communities: A mixed-methods approach. *Social Indicators Research, 56*(1), 43–72.

Phillips, R. (2003). *Community indicators*, PAS report number 517. Chicago: American Planning Association.

Phillips, R., & Pittman, R. (2009). *Introduction to community development*. London: Routledge.

Roberts, S., & Tribe, J. (2008). Sustainability indicators for small tourism enterprises – An exploratory perspective. *Journal of Sustainable Tourism, 16*(5), 575–594.

Sirgy, M. J., Rahtz, D. R., Muris, C., & Underwood, R. A. (1998). Method for assessing resident satisfaction with community-based services: A quality-of-life perspective. *Social Indicators Research, 49*, 279–316.

Urtasun, A., & Gutiérrez, I. (2006). Tourism agglomeration and its impact on social welfare: An empirical approach to the Spanish case. *Tourism Management, 27*(5), 901–912.

Wendel-Vos, G. C. W., Schuit, A. J., Tijhuis, M. A. R., & Kromhout, D. (2004). Leisure time physical activity and health-related quality of life and longitudinal associations. *Quality of Life Research, 13*, 667–677.

Wong, C. (2006). *Indicators for urban and regional planning*. London: Routledge.

World Tourism Organization. (2004). *Indicators of sustainable development for tourism destinations*. Madrid: World Tourism Organization.

Chapter 2
Indicators and Standards in Parks and Outdoor Recreation

Robert E. Manning

Introduction

Management of parks and outdoor recreation is a specific form of the broad field of environmental management. Emerging concepts in environmental management include ecosystem management, sustainability, adaptive management, and indicators and standards. Management of parks and outdoor recreation was an early adopter of all of these concepts, particularly indicators and standards.

Contemporary parks and outdoor recreation management frameworks are built on a procedural foundation of formulating indicators and standards, monitoring indicator variables, and applying management practices to ensure that standards are maintained. Formulation of indicators and standards can be guided by a program of natural and social science research, including application of normative theory and methods. Research has supported formulation of a diverse array of indicators and standards in the national park system and related areas.

Emerging Concepts in Environmental Management

Contemporary environmental management is being guided by a number of emerging concepts, including ecosystem management, sustainability, adaptive management, and indicators and standards. While definitions and operational procedures for these concepts are still evolving, several principles can be isolated that might be broadly applicable to many forms of environmental management, including parks and outdoor recreation. First, environmental management must address the integration of environment and society (Agee & Johnson, 1987; Grumbine, 1994; Society of American Foresters, 1993). The integrity of important ecological processes must be protected, but natural and environmental resources also must be managed for the benefits of society. Thus, the foundational principle of *ecosystem management* has been defined as "regulating... ecosystem structure

R.E. Manning (✉)
University of Vermont, Burlington, VT, USA
e-mail: Robert.Manning@uvm.edu

M. Budruk, R. Phillips (eds.), *Quality-of-Life Community Indicators for Parks, Recreation and Tourism Management*, Social Indicators Research Series 43, DOI 10.1007/978-90-481-9861-0_2, © Springer Science+Business Media B.V. 2011

and function … to achieve socially desirable conditions" (Agee & Johnson, 1987) and "integrating …ecological relationships within a complex sociopolitical and values framework" (Grumbine, 1994).

Second, managing the environment for the benefits of the present generation should not preclude the ability of future generations to attain needed environmentally related benefits. This principle is at the heart of the emerging concept of *sustainability* as originally outlined by the World Commission on Environment and Development (1987) and as sustainability is now being applied in many environmental and related fields.

Third, environmental management should be conducted within a framework that identifies goals and objectives and works toward these ends through a program of monitoring and management. A report by the Ecological Society of America recommends that environmental management be "driven by explicit goals and made adaptable by monitoring and research" (Christensen et al., 1996). This principle is fundamental to the evolving concept of *adaptive management*, which emphasizes the role of ongoing monitoring and evaluation as a way of informing environmental management (Holling, 1978; Lee, 1993; Stankey, Clark, & Bormann, 2005; Walters, 1986).

Fourth, *indicators and standards* are emerging as a substantive focus of environmental management. The contemporary scientific and professional literature contains thousands of references to the expanding use of environmental and social indicators (see, for example, McKenzie, Hyatt, & McDonald, 1992; National Research Council, 2000; Niemi & McDonald, 2004; US Environmental Protection Agency, 2002 for reviews of this literature). There is evidence that early humans relied on environmental indicators such as migratory animal movements for information about changing natural conditions (Niemi & McDonald, 2004). However, modern scientific use of environmental indicators can be traced to the work of Clements (1920), who laid the foundation for the use of plants as indicators of ecological conditions and processes (Morrison, 1986). Perhaps the most widely known early use of environmental indicators is "the canary in the coal mine" as a measure of air quality (Burrell & Siebert, 1916). Environmental indicators have expanded to include a host of measures other than observation of plant and animal species, and they sometimes use indexes comprising multiple variables.

Social indicators also have a relatively long history of use. An early example is the work of Odum (1936), who developed a large suite of indicators of socioeconomic conditions in the southern United States for purposes of regional planning (Force & Machlis, 1997). Economic indicators such as unemployment rate, interest rate, and gross national product (GNP), along with social indicators such as crime rate, literacy, and life expectancy, have been central to economic and social planning in the United States for many years. As noted above, the concept of ecosystem management has emphasized the connections between the environment and society, and this has suggested that environmental management should include indicators of both ecological and associated social conditions.

Indicators and Standards

Contemporary emphasis on indicators (and standards to an increasing degree) in environmental management is a direct outgrowth of the United Nations Conference on Environment and Development (popularly known as the Earth Summit) held in Rio de Janeiro in 1992. This conference prepared a plan of action titled *Agenda 21* to achieve sustainability on a global basis and called for identification of "indicators of sustainable development." The Commission on Sustainable Development was established to help ensure effective follow-up. To monitor the implementation of Agenda 21, the commission established 134 (more recently reduced to 57) broad-ranging indicators (Commission on Sustainable Development, 2001). The list includes environmental (e.g., ambient concentration of air pollutants in urban areas), social (e.g., percent of population with access to safe drinking water), and institutional/managerial (e.g., implementation of national sustainable development strategy) variables.

The work of the Commission on Sustainable Development has been extended to many areas of environmental management by a host of governmental and nongovernmental organizations. For example, one of the more highly developed applications of indicator-based approaches to environmental management is the current program of sustainable forestry. In 1993, following the Earth Summit, the International Seminar of Experts on Sustainable Development of Boreal and Temperate Forests was held in Montreal. A further outgrowth of this initiative was the Working Group on Criteria and Indicators for the Conservation and Sustainable Management of Temperate and Boreal Forests, popularly known as the *Montreal Process Working Group*. In 1995, in its meeting in Santiago, the working group developed 7 criteria and 67 indicators (popularly known as the *Santiago Declaration*) to guide sustainable forestry at the country or national level. The seven criteria are analogous to management objectives. For example, the first criterion is *conservation of biological diversity*. The 67 indicators are measurable, manageable variables that can be used as proxies for these objectives. For example, indicators of the first criterion include the number of forest-dependent species and extent of area by forest type relative to total forest area. The seven criteria included in the Montreal Process range from ecological to social to institutional considerations. The criteria and indicators included in this program are intended to provide a commonly agreed-upon understanding of what is meant by sustainable forest management and to be a mechanism for evaluating a country's success at achieving sustainability at the national level. Given substantive differences among nations regarding basic forest-related conditions (e.g., amount of forest land, population density), standards for indicator variables are left to the discretion of countries that choose to endorse the Santiago Declaration. These countries are expected to monitor indicators on a regular basis, with resulting data suggesting the degree to which sustainability in forest management is being achieved and informing national and international forestry-related policy and management.

Indicators and Standards in Parks and Outdoor Recreation

The field of parks and outdoor recreation was an early adopter of the concept of indicators and standards. Park and outdoor recreation professionals have wrestled for decades with the issue of *carrying capacity*, a special manifestation of sustainability (Shelby & Heberlein, 1986; Manning, 1999; 2007). Carrying capacity addresses the inherent tension between use of parks and related areas and protection of park resources and the quality of the visitor experience. Research demonstrates that outdoor recreation can cause impacts to park resources in the form of trampled vegetation, compacted and eroded soils, water pollution, and disturbance of wildlife (Hammitt & Cole, 1998). Moreover, recreation use can also degrade the quality of the visitor experience in the form of crowding, conflicting uses, and aesthetic implications of the resource impacts noted above (Manning, 1999). How much use and associated impact is acceptable in parks and related areas?

Research and management experience suggest that this and related questions can be answered only as they relate to *management objectives*. For example, what degree of environmental protection should be maintained in a given area and what type of visitor experience should be provided? In the context of parks and outdoor recreation, management objectives are sometimes called *desired conditions* (Manning, 2007; National Park Service, 1997). To make them fully operational, management objectives must ultimately be expressed in quantitative terms as indicators and standards. Indicators are measurable, manageable variables that help define the desired quality of parks and outdoor recreation. As such, they are proxies for management objectives. Standards define the minimum acceptable condition of indicators. Based on this conceptual approach, carrying capacity (or sustainability) of parks and outdoor recreation areas can be defined and managed in an operational way as the level and type of visitor use that can be accommodated without violating standards for relevant indicators.

Indicators and standards have been adopted as the conceptual foundation of contemporary park and outdoor recreation planning and management frameworks, including limits of acceptable change (LAC) (Stankey et al., 1985), visitor impact management (VIM) (Graefe, Kuss, & Vaske, 1990), and visitor experience and resource protection (VERP) (Manning, 2001; National Park Service, 1997). All of these frameworks function through a similar core sequence of steps: (1) formulate management objectives/desired conditions and associated indicators and standards, (2) monitor indicator variables, and (3) apply management practices to ensure that standards are maintained (Manning, 2004). In this way, carrying capacity is defined and parks and outdoor recreation areas are managed in a way that sustains park resources and the quality of the visitor experience.

Research to Support Formulation of Indicators and Standards

Formulation of indicators and standards for parks and outdoor recreation will always require some element of management judgment. However, such judgments should be as informed as possible (Manning & Lawson, 2002). Recent research has identified and tested a number of scientific approaches that can inform development of indicators and standards for parks and outdoor recreation.

Potential indicators can be identified through qualitative interviews with park visitors and other stakeholders. In this process, respondents are asked a series of open-ended questions designed to provide an understanding of what contributes to and detracts from the quality of the visitor experience. Quantitative surveys can also be used. In this process, batteries of close-ended questions are used with defined response scales, and respondents report the importance of a series of potential indicators. Other approaches used to identify indicators include (1) measures of the "salience" of indicators as derived from normative research methods (described in the next section of this chapter), (2) stated choice surveys and related statistical methods, (3) ecological research on the relationship between visitor use and associated impacts, (4) review of pertinent legislation and management agency policy, and (5) public input associated with park and outdoor recreation planning and management.

Research can also help identify potential standards for indicator variables. Normative theory and methods are increasingly used for this purpose (Heberlein, 1977; Manning, 1999; 2007; Shelby & Heberlein, 1986; Shelby, Vaske, & Donnelly, 1996; Vaske, Donnelly, & Shelby, 1992; Vaske, Donnelly, & Shelby, 1993; Vaske, Graefe, Shelby, & Heberlein, 1986). Building on the work of Jackson (1965), park visitors or other stakeholders can be asked to evaluate the acceptability of a range of environmental and social conditions. For example, visitors might be asked to rate the acceptability of encountering an increasing number of recreation groups while hiking along trails. Resulting data would measure the personal crowding norms of each respondent. These data can then be aggregated to test for social crowding norms or the degree to which norms are shared across groups.

Social norms can be illustrated graphically, as shown in Fig. 2.1. Using hypothetical data associated with the example described above, this graph plots average acceptability ratings for encountering increasing numbers of visitor groups along trails. The line plotted in this illustration is sometimes called a social norm curve.

Norm curves like that illustrated in Fig. 2.1 have several potentially important features or characteristics. First, all points along the curve above the neutral line – the point on the vertical axis where evaluation ratings fall from the acceptable into the unacceptable range – define the "range of acceptable conditions." All of the conditions represented in this range are judged to meet some level of acceptability by about half of all respondents. The "optimum condition" is defined by the highest point on the norm curve. This is the condition that received the highest rating of acceptability from the sample as a whole. The "minimum acceptable condition" is defined as the point at which the norm curve crosses the neutral line. This is the condition that approximately half of the sample finds acceptable and half finds unacceptable. Norm "intensity" or norm "salience" – the strength of respondents' feelings about the importance of a potential indicator of quality – is suggested by the distance of the norm curve above and below the neutral line. The greater this distance, the more strongly respondents feel about the indicator of quality or the condition being measured. High measures of norm intensity or salience suggest that a variable may be a good indicator because respondents feel it is important in defining

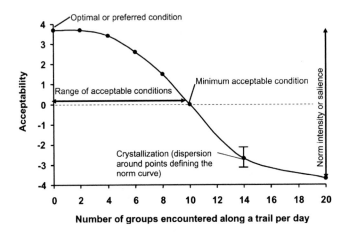

Fig. 2.1 Hypothetical social norm curve

the quality of the recreation experience. "Crystallization" of the norm concerns the amount of agreement or consensus about the norm. It is usually measured by standard deviations or other measures of variance of the points which describe the norm curve. The less variance or dispersion of data around those points, the more consensus there is about social norms. Norm curves are sometimes constructed with the vertical axis of the graph representing the percentage of respondents who report each level of impact as the maximum acceptable.

Norms can also be measured using a shorter, open-ended question format by asking respondents to report the maximum level of impact that is acceptable to them. In the example illustrated in Fig. 2.1, respondents would simply be asked to report the maximum number of groups they would find acceptable to meet while hiking along trails during a day's time. This format is designed to be less burdensome to respondents, but it also yields less information.

Park and Outdoor Recreation Indicators and Standards

Indicators and standards have been formulated for many park and outdoor recreation areas, and much of this work has been supported by a program of natural and social science (Manning, 2007; 2009). For example, interviews were conducted with visitors and other stakeholders in conjunction with development of a management plan for Arches National Park, Utah. The purpose of the interviews was to help identify indicators for the visitor experience, including the ways in which the condition of park resources affected the visitor experience. A semi-structured interview script was developed that asked a series of probing, open-ended questions about what respondents felt were the most important qualities or characteristics of the visitor experience at Arches. Interviews were conducted in the park with 112 visitors at 7 sites. In addition, 10 focus group sessions were also conducted with a total of 83 participants. Participants included park staff, visitors who participated in the park's interpretive programs, and residents of the local community.

Responses were initially coded into 91 categories and then grouped into several major themes or subject matter classes. Themes that best met the characteristics of good indicators (e.g., measurable, manageable, related to visitor use, important in affecting the quality of the visitor experience) included crowding at attraction sites and damage to soils and vegetation caused by visitors walking off designated trails.

A second phase of research was conducted to help inform development of standards for these indicator variables. These studies used visual research methods to portray a range of social and environmental conditions in the park (Manning & Freimund, 2004; Manning, Lime, Freimund, & Pitt, 1996). A series of 16 computer-edited photographs were prepared showing a range of visitors at Delicate Arch, a principal park attraction site. Representative photographs are shown in Fig. 2.2. The number of visitors in the 16 study photographs ranged from 0 to 108, with the upper end of the range designed to show approximately 30% more visitors than the current maximum. The purpose was to illustrate a full range of density conditions, including the near-term future. A representative sample of visitors who had just completed their hike to Delicate Arch was asked to examine the photographs in random order and rate the acceptability of each on a scale that ranged from –4 ("very unacceptable") to +4 ("very acceptable"), with a neutral point of 0. Individual acceptability ratings were aggregated into a social norm curve and this provided an empirical foundation for helping to formulate a crowding-related standard for this site. The social norm curve for crowding at Delicate Arch (based on regression analysis of

Fig. 2.2 Representative study photographs showing a range of visitor use levels at Delicate Arch

Fig. 2.3 Social norm curve
for crowding at Delicate Arch

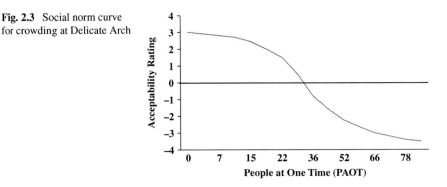

resulting data) is shown in Fig. 2.3. The social norm curve crossed the neutral point
of the acceptability scale (i.e., falls out of the acceptable range and into the unacceptable range) at 30 people at one time, and this represents a minimum acceptable
standard.

A similar approach was taken in measuring visitor-based standards for trail
impacts. A program of ecological research demonstrated the ways in which soils and
vegetation were impacted when visitors walked off designated trails, contributing to
the creation of social or visitor-caused trails. This pattern of resource impact was
represented in a series of computer-edited photographs as shown in Fig. 2.4. Park
visitors were asked to rate the acceptability of these photographs using the response
scale described above. The resulting social norm curve is shown in Fig. 2.5. It is
clear that visitors are very sensitive to these types of resource impacts suggesting

Fig. 2.4 Study photographs showing a range of trail impacts at Arches National Park

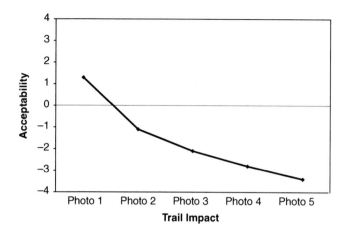

Fig. 2.5 Social norm curve for trail impacts at Arches National Park

that park management should work hard to encourage hikers to stay on designated trails and that the standard for trail impacts should be quite low. Findings from the studies outlined above were incorporated into a management plan for Arches National Park (National Park Service, 1995).

A program of natural and social science research conducted by the Park Studies Laboratory at the University of Vermont, USA, and its collaborators has designed and applied similar studies at nearly 30 diverse units of the US national park system (Manning, 2007). These studies have identified an array of park and outdoor recreation-related indicators as shown in Table 2.1. These indicators represent the resource, social, and managerial dimensions of parks and outdoor recreation. Data to support formulation of standards for these indicators have also been developed.

Table 2.1 Indicators of quality for selected US national parks

PAOT at attraction sites
PAOT on trails
Trail encounters
Campsite encounters
Trail impacts
Campsite impacts
Social trails
Traffic congestion
River encounters
Recreation conflict
Waiting times
Litter
Graffiti
Tour group size/duration/frequency
Trail development
Campsite development
Noise

Conclusion

Indicators and standards are emerging as an important conceptual formulation in parks and outdoor recreation and environmental management more broadly. There are obvious parallels between the concept of carrying capacity as conventionally applied in the field of parks and outdoor recreation and the more contemporary notion of sustainability in environmental management: both address the inherent tension between use of the environment and protection of its basic integrity and related values. Indicators and standards offer a quantitative and operational way to analyze, define, and manage these issues. Indicators and standards also address the integration of the environment and society – the natural and social sciences – as suggested by the concept of ecosystem management. Indicators and standards can (and should) be formulated for both environmental resources and values they represent to people. Moreover, societal values and associated norms can be instrumental in helping to identify potential indicators and associated standards. Finally, indicators and standards help facilitate the concept of adaptive management. Indicators provide the focus for a long-term program of monitoring, and resulting data offer important guidance about conditions and trends in environmental quality and associated values and the effectiveness of management programs.

Application of indicators and standards can be supported and informed through a program of research. This work has been applied broadly in the field of parks and outdoor recreation and has identified a range of indicators and standards for both resource and social conditions across a spectrum of parks and outdoor recreation areas. In this way, the field of parks and outdoor recreation may be a leader in environmental management more broadly.

References

Agee, J., & Johnson, D. (1987). *Ecosystem management for parks and wilderness*. Seattle, WA: University of Washington Press.

Burrell, G., & Siebert, F. (1916). Gases found in coal mines. *Miners circular 14*. Washington, DC: Bureau of Mines, US Department of Interior.

Christensen, N., Bartuska, A., Brown, J., Carpenter, S., D'Antonio, C., Francis, R., et al. (1996). Report of the ecological society of America committee on the scientific basis for ecosystem management. *Ecological Applications, 6*, 665–691.

Clements, F. (1920). *Plant indicators*. Washington, DC: Carnegie Institute.

Commission on Sustainable Development. (2001). *Indicators of sustainable development: Guidelines and methodologies*. New York: United National Division for Sustainable Development.

Force, J., & Machlis, G. (1997). The human ecosystem, Part II: Social indicators in ecosystem management. *Society and Natural Resources, 10*, 360–382.

Graefe, A., Kuss, F., & Vaske, J. (1990). *Visitor impact management: The planning framework*. Washington, DC: National Parks and Conservation Association.

Grumbine, R. (1994). What is ecosystem management? *Conservation Biology, 8*, 27–38.

Hammitt, W., & Cole, D. (1998). *Wildland recreation: Ecology and management*. New York: Wiley.

Heberlein, T. (1977). Density, crowding, and satisfaction: Sociological studies for determining carrying capacities. *Proceedings: River Recreation Management and Research Symposium.* USDA Forest Service General Technical Report, NC-28, 67–76.

Holling, C. (1978). *Adaptive environmental assessment and management.* London: Wiley.

Jackson, J. M. (1965). Structural characteristics of norms. In I. D. Steiner, & M. F. Fishbein (eds.), *Current studies in social psychology* (301–309). New York: Holt, Rinehart & Winston.

Lee, K. (1993). *Compass and gyroscope: Integrating science and politics for the environment.* Washington, DC: Island Press.

Manning, R. (1999). *Studies in outdoor recreation: Search and research for satisfaction* (2nd ed). Corvallis, OR: Oregon State University Press.

Manning, R. (2001). Visitor experience and resource protection: A framework for managing the carrying capacity of national parks. *Journal of Park and Recreation Administration, 19,* 93–108.

Manning, R. (2004). Recreation planning frameworks. *Society and natural resources: A summary of knowledge* (83–93). Jefferson, MO: Modern Litho.

Manning, R. (2007). *Parks and carrying capacity: Commons without tragedy.* Washington, DC: Island Press.

Manning, R. (2009). *Parks and people: Managing outdoor recreation at Acadia National Park.* Lebanon, NH: University Press of New England.

Manning, R., & Freimund, W. (2004). Use of visual research methods to measure standards of quality for parks and outdoor recreation. *Journal of Leisure Research, 36*(4), 552–579.

Manning, R., & Lawson, S. (2002). Carrying capacity as "informed judgment": The values of science and the science of values. *Environmental Management, 30,* 157–168.

Manning, R., Lime, D., Freimund, W., & Pitt, D. (1996). Crowding norms at frontcountry sites: A visual approach to setting standards of quality. *Leisure Sciences, 18,* 39–59.

McKenzie, D., Hyatt, D., & McDonald, V. (1992). *Ecological indicators* (Vols. 1 and 2). New York: Elsevier Applied Science.

Morrison, M. (1986). Bird populations as indicators of environmental change. *Current Ornithology.* New York: Plenum.

National Park Service. (1995). *VERP implementation for Arches National Park.* Denver, CO: US National Park Service.

National Park Service. (1997). *VERP: The visitor experience and resource protection (VERP) framework—A handbook for planners and managers.* Denver, CO: Denver Service Center.

National Research Council. (2000). *Ecological indicators for the nation.* Washington, DC: National Academy Press.

Niemi, G., & McDonald, M. (2004). Application of ecological indicators. *Annual Review of Ecology, Evolution, and Systematics, 35,* 89–111.

Odum, H. (1936). *Southern regions of the United States.* Chapel Hill, NC: University of North Carolina Press.

Shelby, B., & Heberlein, T. (1986). *Carrying capacity in recreation settings.* Corvallis, OR: Oregon State University Press.

Shelby, B., Vaske, J., & Donnelly, M. (1996). Norms, standards and natural resources. *Leisure Sciences, 18,* 103–123.

Society of American Foresters. (1993). *Sustaining long-term forest health and productivity.* Bethesda, MD: Society of American Foresters.

Stankey, G., Clark, R., & Bormann, B. (2005). *Adaptive management of natural resources: Theory, concepts, and management institutions.* USDA Forest Service General Technical Report, PNW-GTR–654.

Stankey, G., Cole, D., Lucas, R., Peterson, M., Frissell, S., & Washburne, R. (1985). *The limits of acceptable change (LAC) system for wilderness planning.* USDA Forest Service General Technical Report, INT–176.

US Environmental Protection Agency. (2002). *A SAB report: A framework for assessing and reporting on ecological conditions.* Washington, DC: US Environmental Protection Agency.

Vaske, J., Graefe, A., Shelby, B., & Heberlein, T. (1986). Backcountry encounter norms: Theory, method, and empirical evidence. *Journal of Leisure Research, 18,* 137–153.

Vaske, J., Donnelly, M., & Shelby, B. (1992). Establishing management standards: Selected exam-
ples of the normative approach. *Defining wilderness quality: The role of standards in wilderness
management – A workshop proceedings*. USDA Forest Service General Technical Report,
PNW-305, 23–37.
Vaske, J., Donnelly, M., & Shelby, B. (1993). Establishing management standards: Selected
examples of the normative approach. *Environmental Management, 17*, 629–643.
Walters, C. (1986). *Adaptive management of renewable natural resources*. New York: Macmillan.
World Commission on Environment and Development. (1987). *Our common future*. New York:
Oxford University Press.

Chapter 3
Leisure and Relationship to Quality-of-Life Satisfaction

Ariel Rodríguez

> ... *if we do not know the nature of meaning in leisure and what it entails as a concept and a construct, how can we compare it against, or test its relationship to, other domains?*
>
> (Ragheb, 1996, p. 245)

Introduction

The quality of a person's life is often said to be influenced directly or indirectly by a number of factors. Among these, for example, are a person's health, their employment, their friends and family, and their leisure (Hagerty et al., 2001). But what exactly is meant by the term leisure? If one were to look up the definition(s) of leisure, it might include a number of terms including time, obligations, opportunity, relaxation, ease, pleasure, experience, autonomy, intrinsic motivation, intrinsic reward, social, control, and even freedom. From this list, one can glean that leisure is used to describe a number of different meanings. While these (and other) descriptors might help us to understand how the term leisure is more commonly used, it does little to help researchers develop theoretical and conceptual models to better understand the relationship between leisure and those constructs that are important in maintaining or improving the quality of our lives. In other words, the issue raised by Ragheb (1996) in the introductory quote has yet to be fully addressed which has implications to our understanding of the relationship between leisure and life satisfaction.

There are a few issues with using the term leisure to describe different meanings. First and foremost, using one word to describe different meanings creates confusion. Aside from literally asking a person what they mean when they are using the term leisure, it is a guessing game. As an example, in a study by Shin and Rutkowski (2003), a national sample of Koreans were asked to indicate their level of satisfaction

A. Rodríguez (✉)
Arizona State University, Phoenix, AZ, USA
e-mail: Ariel.Rodriguez@asu.edu

M. Budruk, R. Phillips (eds.), *Quality-of-Life Community Indicators for Parks, Recreation and Tourism Management*, Social Indicators Research Series 43, DOI 10.1007/978-90-481-9861-0_3, © Springer Science+Business Media B.V. 2011

or dissatisfaction with a number of life domains including health, income, education, and leisure and recreation. Yet, it is not clear how study participants interpreted the term leisure in their study, thus it is not apparent with what they were satisfied or dissatisfied.

Second, different elements of leisure have positive impacts on different components of our life satisfaction, the cognitive component of subjective well-being (Diener, Suh, Lucas, & Smith, 1999), while others have negative impacts. For example, if one were to look at leisure as an activity done during one's free time, it would be difficult to argue that using drugs such as cocaine or heroin would contribute positively to one's life satisfaction. On the other hand, going for a jog has been found to positively influence a person's life satisfaction (Rodríguez, Látková, & Sun, 2008). Combining all the elements together under the label of "leisure" makes it difficult to understand the true complex relationships between leisure and our life satisfaction.

Lastly, and an off-shoot of the second point, our theoretical understanding of human behavior and therefore the outcomes of these behaviors is often limited by our ability to accurately conceptualize and ultimately operationalize a construct. Several leading leisure scholars have made attempts to come to an agreed-upon meaning of leisure over the past 40 years only to have these efforts brushed aside to the graveyard of mounting leisure definitions. Therefore, I will not focus on providing yet another "definitive" definition of leisure, but will instead focus on the more contemporary conceptions or meanings of leisure.

The term leisure has been used to identify three primary meanings: time, an activity, and an experience (Mannell & Kleiber, 1997). As time, leisure refers to the discretionary time that we have, time that is free from obligations. As an activity, leisure is defined as the activities that we do during our free time. These activities may be done for positive reasons, such as for rejuvenation purposes, or for negative purposes, such as to forget about our problems as may be done through heavy drinking. Finally, as an experience, leisure refers to participating in leisure activities for intrinsically motivated reasons. If one were looking at a target, leisure experience would be the innermost circle encompassed by leisure activity which is encompassed by leisure time. In other words, leisure time provides the basis for both leisure activity and leisure experiences. Moreover, using this rationale, all leisure experiences result from participation in leisure activities, but not all leisure activity participation results in a leisure experience. Given the different meanings of leisure, the question that begs to be asked is whether or not leisure, conceptualized as either time, an activity, or an experience, differs in its relationship with life satisfaction. The remainder of this chapter is dedicated to answering this question.

Leisure as Free Time

Over the past five decades, between 1965 and 2003, the amount of leisure time has increased for both women and men in the United States. For women, it increased between 4 and 8 h/week. This was largely caused by a decline in home production work hours. As for men, their leisure time increased between 6 and 9 h. This was due predominantly because of a decline in market work hours (Aguiar & Hurst, 2007). Yet, despite these findings, life satisfaction measures within various countries show consistent scores over time (Diener, Oishi, & Lucas, 2003).

When comparing between ten more developed countries (e.g., Australia, Denmark, Sweden, the UK, and the USA) and genders, studies have found that those in Denmark and the Netherlands tend to have more free time than those in other countries. In fact, they are the only countries from those compared where both men and women have more than 40 h/week of leisure time (Bittman & Wajcman, 2000). Despite this, cross-cultural studies on subjective well-being have found that individuals in either Denmark or the Netherlands do not top the list of highest levels of subjective well-being (Diener, Diener, & Diener, 1995). In the USA, women have about an hour and a half of more free time than men whereas in Italy, men have about 6 more hours of free time than women. Among heterosexual couples who are married, men have approximately 40 more minutes of free time per week than women in the USA, and in Canada, men have a little over 5 h more free time per week than women (Bittman & Wajcman, 2000). Similar to differences between countries, gender differences do not seem to significantly influence how satisfied a person is with their life (Diener et al., 1999).

Leisure as free time or simply leisure time looks at the concept of leisure solely from the perspective of time. In the literature, leisure time is often synonymous with discretionary time or time that is available for us to do as we please after we have fulfilled our employment duties and other obligations (Russell, 2009); it is time at one's disposal (Bittman & Wajcman, 2000). But in order to further understand leisure time and our current understanding of its relationship with life satisfaction, we must first understand what time is and how we make sense of it? According to Page-Wood and Lane (1989), there are at least four ways of understanding time orientation of individuals: economic, socio-cultural, psychological, and measurement.

From an economic perspective, time is looked at as a commodity with potential utility. As such, it suggests that individuals are involved in productive activities throughout their lives. From this perspective, leisure time is a "remainder" category or time which is spent not being productive (Page-Wood & Lane, 1989). In other words, leisure time tends to be synonymous with unproductive time. Of the four perspectives of time, the economic perspective has predominantly guided studies analyzing leisure time.

From a socio-cultural perspective, time is interpreted by socio-cultural factors within a specific environment (Page-Wood & Lane, 1989). For example, several ancient cultures such as the Mayan, Hopi, and Greek understood time from a cyclical perspective (Russell, 2009). Time was reoccurring and a normal part of nature. Therefore, leisure time is never wasted as it renews itself each day. There is no need to control time or to save time since it comes back every day.

When we focus on how individuals perceive time from a personal perspective, we are analyzing time from a psychological perspective. From this perspective, we can understand individual differences and relationships that we have with time.

Additionally, we can better understand the extent to which individuals are satisfied with their leisure time (Page-Wood & Lane, 1989).

To understand the last perspective, measurement perspective, one simply has to look at their watch, clock, or cell phone if that is what they use to tell time. This perspective refers to our physical time such as hours, minutes, and seconds. It also includes days, months, and years. In a measurement perspective, time can be measured, thus it can have parameters (Page-Wood & Lane, 1989). For example, if I want to be able to spend 1 h at the gym today, I need to leave now so that I can make it back on time to shower and have dinner with my family. It is this perspective of leisure time that is often inferred when issues of time scarcity, paucity of leisure time, or harried leisure are discussed.

Measures of Leisure Time

Leisure time has been operationalized in a number of ways throughout the literature, yet the most common method, which is more closely linked to the economic perspective of time, is to measure it objectively as time left over after one has fulfilled their work and other obligatory duties. There are a few modifications to this basic structure, but the principle is the same. For example, Aguiar and Hurst (2007) identified four incremental ways of measuring leisure time: (1) summing together all time spent on "entertainment/social activities/relaxing" and "active recreation"; (2) all time in the first measure plus time spent sleeping, eating, and on personal care; (3) all time in the second measure plus time spent in child care; and (4) all time that is the residual of total work.

While these measures have provided the basis for important empirical research, they may be hindered by at least three possible issues. First, this measure often assumes individuals have a monochronic manner of processing time as opposed to a polychronic. The difference is that from a monochronic manner of processing time, a person does one thing at a time; people who process time from a polychronic perspective do multiple activities at the same time; they are multitaskers (Page-Wood & Lane, 1989).

Second, it does not differentiate between obligated activities and committed activities. Because we live in a society where space and time need to be reserved for various leisure activities, this reservation might constitute a level of commitment necessary to participate in the specific leisure activity. But does this commitment necessitate obligation given that once we have committed to participate in an activity, are we not now to a certain extent obligated to do so? Given the social nature of leisure activities, this amount of time might cloud our understanding of what constitutes true leisure time. A possible solution would be to classify time using six kinds of uses: income producing obligated, income producing not obligated, committed obligated, committed not obligated, uncommitted unplanned, and uncommitted planned (Lane & Lindquist, 1988).

Lastly, these measures assume a constant in how time is spent during leisure time, the experiences individuals have during their leisure time, and the outcomes obtained during one's leisure time. Given the variety of ways individuals spend their leisure time, measures of leisure time outlined in this chapter may reach erroneous conclusions about the relationship between leisure time and life satisfaction.

Other measures that take into consideration different perspectives have also been developed. Two measures in particular, which are more aligned with psychological and measurement perspectives of time, are the experience sampling methodology (ESM) and the day reconstruction method (DRM). It is important to note that while these measures may be used to assess leisure time, they may also be used to assess different outcomes related to activities, such as leisure activities, done during one's leisure time.

Of the two techniques identified here, ESM is the more established. ESM simply refers to a method of collecting data where individuals are in their natural environment (i.e., not in a laboratory). Individuals receive cues at different times throughout the day to provide information about various questions such as level of happiness or current mood. Cues come in various forms, but more common cues focus on devices such as portable electronic devices such as cell phones, hand-held computers such as a Blackberry, pagers, and even watches. According to Scollon, Kim-Prieto and Diener (2003), "Much of its [ESM] popularity can be attributed to its ability to delve beyond single-time self-report measurement to answer complex questions about lives... as well as its ability to provide solutions to nagging methodological problems, such as memory biases" (p. 5). Today, ESM has been used in hundreds of studies over the past 30 years. Yet, while ESM has many positive attributes such as the reduction of memory bias, is great with multiple-methods, and allows researchers to intensively investigate within-person processes, it has a few weaknesses which include sample limitations (it is not for every group of individuals), selection bias and attrition (given the intensity of the study), and potential lack of motivation issues with participants (again, attributed to the intensity of the study) (Scollon et al., 2003). For a more comprehensive discussion of ESM, readers are referred to Scollon et al. (2003).

The day reconstruction method (DRM) is a relatively newer technique which combines features of time-budget measurement and experience sampling (Kahneman, Krueger, Schkade, Schwarz, & Stone, 2004b). The purpose of DRM is to describe experiences a person has on a given day in an efficient manner. To do this, study participants are asked to answer a structured self-administered questionnaire which systematically reconstructs their previous day. When compared to ESM, Kahneman et al. (2004b) feel DRM is more efficient than ESM as, "it imposes less respondent burden; does not disrupt normal activities; and provides an assessment of contiguous episodes over a full day, rather than a sampling of moments" (p. 1776). Additionally, DRM provides time-budget information which tends not to be collected as effectively with experience sampling (Kahneman et al., 2004b). For more information about DRM, readers are encouraged to review the day reconstruction method instrument documentation (Kahneman, Krueger, Schkade, Schwarz, & Stone, 2004a). While DRM shows promise, additional research is needed to better understand its strengths and weaknesses.

Quality-of-Life Indexes for Leisure Time

There are a number of quality-of-life indexes which include leisure time items (Hagerty et al., 2001), but they tend to fall under either the economic or psychological perspectives of time. For example, the World Health Organization Quality of Life–100 (WHOQOL–100) asks, "How much do you enjoy your free time?" (World Health Organization, 2000a). This item is consistent with a more psychological perspective of time attempting to identify a level of satisfaction with individual leisure time. The American Demographics Index of Well-Being took a different approach and measured the amount of leisure time as 168 (total hours in a week) minus the number of weekly hours worked (Hagerty et al., 2001). This measure is consistent with the economic perspective of time and is also consistent with the fourth measure of leisure time aforementioned from Aguiar and Hurst (2007).

Leisure Time and Life Satisfaction

Given that perspectives and therefore measures of leisure time vary considerably, it is not too much of a surprise that there are differences in the relationship between leisure time and life satisfaction based on these techniques. Specifically, studies that have used methods more consistent with the economic perspective of time tend to find that leisure time explains little to no variance in life satisfaction, subjective well-being, or quality of life. As identified in the beginning of the leisure time section of this chapter, increases in the amount of leisure time has not translated into increases in life satisfaction or subjective well-being measures.

On the other hand, studies that have used techniques more consistent with psychological perspectives of time, such as those that have used DRM or ESM, have been able to provide more comprehensive information on leisure time. For instance, Kahneman et al. (2004b) were able to identify the amount of time individuals spent on a variety of activities including leisure activities and connect this with the level of positive and negative affect participants felt during their participation (affect results presented in leisure activity section of this chapter). In a convenience sample of 1,018 employed women, they found that women spent 0.2 mean hours of their day exercising and being in intimate relations, while they spent the largest portion of their day working 6.9 mean hours/day. Additionally, the women in the sample spent 2.6 mean hours a day with friends and 3.4 mean hours a day alone (Kahneman et al., 2004b). Using ESM, Csikszentmihalyi and Hunter (2003) were able to assess which day of the week and time of the day individuals ($n = 828$ primary school students) were happiest. They found that on Mondays, individuals seemed to be at their lowest levels of happiness while on Saturdays, they were the happiest. Additionally, they were happiest around noon–1:30 (lunch time) and from 3 to 4:30 (when many students get out of school and have free time or attend after-school programs) (Csikszentmihalyi & Hunter, 2003).

Leisure as an Activity

Most of the empirical research that is available on the relationship between leisure and life satisfaction falls under leisure as an activity. As indicated earlier, leisure as an activity simply refers to activities individuals do during their free time. Thus, if a person enjoys hitting a few rounds of golf during their free time, then "playing golf" would be the leisure activity. Similarly, if a person enjoys watching television from dusk until dawn on their Saturday or going to mass on Sunday morning, these too would be considered leisure activities.

Measures of Leisure Activity

As mentioned earlier in this chapter, leisure activities may be analyzed using techniques such as experimental sampling methodology (ESM) or day reconstruction method (DRM). These provide more comprehensive measurement techniques to analyzing leisure activities. Yet, most studies that fall under this meaning of leisure tend to provide study participants with a list of activities or a list of categories from which to choose from. Study participants then indicate to what extent (i.e., numerical frequency or in a Likert-type scale such as from "do not participate" to "participate often") they participate in each of the activities or categories within a time frame (e.g., past 4 weeks) predetermined by the investigator. An example activity category might be outdoor-physical activities which would include fishing or hunting, working in the garden, and going on walks (Silverstein & Parker, 2002).

The rationale for which activities or categories are chosen tends to differ between studies. One primary reason for their selection has to do with the study participants. Specifically, what activities or categories do the study participants actually participate in? There are some activities that are unique to specific cultures and geographic locations. For instance, a leisure activity survey done in the USA would probably not include sepak takraw, as there would be relatively minimal participants compared to other activities such as baseball, skiing, or even playing chess. On the other hand, if the question was posed in countries throughout Southeast Asia, there would be a much larger number of individuals who participate in this activity.

Differences between specific groups of individuals such as between men and women or those who have higher incomes and those who do not also influence the activity selection process. It is often difficult to decide which activities may be more appropriate for which group of individuals without first obtaining information from the group. For example, in a study of professional women, a list was not available of activities that professional women commonly do, thus the researchers conducted a pilot study to ascertain these activities (Lewis & Borders, 1995).

Season, including weather, also plays a role in which activities are selected. For example, asking study participants about categories during the inappropriate season, such as asking about traditionally winter activities during the summer, would make it seem as though few in the sample participate in winter activities. Similarly, asking whether participants hunt when it is not hunting season would also skew the actual number of hunters within study participants.

Quality-of-Life Indexes for Leisure Activity

There are a number of quality-of-life indexes which include leisure activity items (Hagerty et al., 2001). For example, the World Health Organization Quality of Life–100 (WHOQOL–100) asks "To what extent do you have the opportunity for leisure activities?" and "How much are you able to relax and enjoy yourself?" (World Health Organization, 2000a). The WHOQOL-BREF (short version of the WHOQOL–100) only includes the former item (World Health Organization, 2000b). In the Comprehensive Quality of Life Scale (ComQol) (Cummins, 1997), study participants are asked, "Below is a list of leisure activities. Indicate how often in an average month you attend or do each one for your enjoyment (not employment)." The list of leisure activities includes several activities including, but not limited to eating out, visiting family or friend, playing a sport or going to the gym. Similarly, the Netherland Living Conditions Index (LCI) asks about leisure activities (i.e., number of hobbies, number of nondomestic entertainment activities, organizational memberships) and sport activity (i.e., number of times sporting a week and number of sports) (Boelhouwer & Stoop, 1999).

Leisure Activity and Life Satisfaction

Participation in general activities is often considered positively related to life satisfaction (Lloyd & Auld, 2002). This has been found in various samples including professional women (Lewis & Borders, 1995), older adults (Ray, 1979; Riddick, 1985; Silverstein & Parker, 2002), and general adults (Schnohr, Kristensen, Prescott, & Scharling, 2005). For example, in a random sample of Copenhagen residents ($n = 12{,}028$), it was found that individuals who participated in demanding physical activities (e.g., jogging) were less prone to stress and life dissatisfaction than individuals who lived sedentary lifestyles (Schnohr et al., 2005). As described by Hills and Argyle (1998), "Leisure activities are voluntarily undertaken, therefore it is to be expected that individuals participate in them for enjoyment, even when these activities are physically punishing... and that underlying this enjoyment are the positive moods or emotions which the activities generate" (p. 523).

   Yet, while there are some activities that contribute to overall life satisfaction, the amount of life satisfaction variance they explain tends to be relatively small (Rodríguez et al., 2008). For example, in a recent study comparing psychological needs and leisure activities, it was found that the leisure activities analyzed only explained 4% of the variance in life satisfaction when controlling for satisfaction with various psychological needs (Rodríguez et al., 2008). This discrepancy in the relationship between different activities and life satisfaction is not new. Lemon, Bengtson, and Peterson (1972) identified how only informal leisure activities influenced a person's life satisfaction as opposed to formal activities. Other recent studies have found that activity participation did not significantly influence a person's life satisfaction or overall quality of life (Baker & Palmer, 2006; Michalos, 2003). Finally, in a sample of professional women, leisure activities were significant predicators of life satisfaction, but all activities combined only explained 15% of the variance in life satisfaction where job satisfaction, sexual satisfaction, internal locus of control, and life circumstances accounted for 47% of the variance in life satisfaction (Lewis & Borders, 1995).

These findings are somewhat inconsistent with theories used to explain the relationship between leisure activities and life satisfaction, namely, activity theory. According to activity theory, the amount or frequency of activity participation and the degree of intimacy associated with an activity both influence a person's life satisfaction (Lemon et al., 1972). Specifically, the findings may be partly inconsistent with activity theory given that studies predominantly focused on the frequency of activity participation and not in the level of intimacy associated with an activity. A person may do an activity several times without having a high level of intimacy with the activity. For instance, a person may go to church to appease their family or go to the park so that their child can play but have no real connection to either going to church or to the park.

Additionally, these outcomes are only somewhat inconsistent in that there are some activities that by simply participating, individuals can increase their life satisfaction. The study by Rodríguez et al. (2008) showed how participation in physical activity can have a positive outcome on a person's life satisfaction. On the other hand, the study also showed that participation in certain activities can contribute negatively to a person's life satisfaction. Given that most studies which focus on leisure activities tend to focus on naturally positive activities (e.g., going to movies, dancing, reading, camping, hiking, playing bingo), our understanding of traditionally negative activities (e.g., heavy drinking, doing drugs, fighting, masturbating, going to strip clubs, promiscuous and casual sex, orgies, stealing) and how they affect a person's life satisfaction tends to be limited. For instance, if a person enjoys going to strip clubs will this negatively or positively influence their life satisfaction?

There are at least three reasons for why leisure activities often explain little to no variance in life satisfaction, and the first two are due to measurement limitations of focusing solely on the extent of leisure activity participation. The first reason revolves around the condition of leisure activity participation. Specifically, was the activity participated in for intrinsically motivated reasons? If participation was extrinsically motivated, we can expect different degrees of positive outcome (Deci & Ryan, 1987; Ryan & Deci, 2000). Did the participant feel competent with their activity participation? If they did not feel competent, we can also expect reduced levels of positive outcomes (Bandura, 1977; Ryan & Deci, 2000). Finally, was the activity done with close friends or family, did it encourage social interactions? Research has also shown that the promotion of social interactions in our daily activities, including leisure activities, helps to increase a number of positive outcomes including happiness (Clemente & Sauer, 1976; Diener & Seligman, 2002; Ryan & Deci, 2000).

The second reason focuses on the outcomes of the leisure activity. Specifically, understanding the outcome of an activity, any activity, is difficult to do simply by asking a person how many times they participate in the activity. This method of measurement assumes a person has the same outcomes or similar types of outcomes each time they participate. Moreover, because frequencies are used, it further assumes that these outcomes are incremental. Thus, for example, if we assume that a person gets x amount of positive outcome y each time they go to the gym, we would assume that each time the person goes to the gym this y outcome would consistently increase. Anyone who has been to the gym (or who has participated in most any other leisure activity for that matter) understands first hand that sometimes we do different activities in the gym as with other leisure activities which promote differing increases in levels of positive outcome. Additionally, some days we feel great, but others we feel as if we are not really getting physically stronger or are increasing our stamina. Many things contribute to the outcomes of leisure activities which may not have to do with the actual leisure activities such as nutrition and rest, for example. Therefore, we may obtain different levels of outcomes each time we participate in a leisure activity. Concurrently, we might have differing levels of satisfaction with our participation outcomes.

Studies that have tested both leisure activity participation and satisfaction with outcomes of leisure participation (i.e., leisure satisfaction) have found that leisure satisfaction tends to explain more variance in life satisfaction than leisure activity participation. For example, in a study of 565 persons over 55 years old, it was found that leisure satisfaction explained 20% of the variance in life satisfaction while leisure activity participation explained less than 1% of the variance (Ragheb & Griffith, 1982). Similarly, in a study of middle and later life adults, leisure satisfaction explained 16% of the variance in life satisfaction while leisure activity participation did not significantly predict life satisfaction when tested with other variables (Sneegas, 1986).

A possible solution to better understanding outcomes of activities is to use measurement techniques more consistent with ESM and DRM. For example, Kahneman et al. (2004b) were able to differentiate positive and negative affective ratings for different activities done throughout the day. They found that employed women had the highest level of positive affect with intimate relations followed by socializing and relaxing. At the bottom of the list with levels of positive affect were activities such as doing housework, working, and commuting. Additionally, they had the highest level of positive affect when doing activities with friends and the lowest level when being alone (Kahneman et al., 2004b). Similarly, in their study of primary school students, Csikszentmihalyi and Hunter (2003) were able to assess levels of happiness with activities students most often participated in during the day. They found that students felt happiest when talking with friends, followed by eating and watching television. School-related activities such as listening to lectures and doing homework tended to make the bottom of the happiness list (Csikszentmihalyi & Hunter, 2003).

In addition to assuming that individuals have incremental outcomes, when we cluster leisure activities based on factor analysis of participation frequencies (most common technique for developing a classification system of leisure activities), the frequency of participation provides little information about the unique outcomes of each activity within the respective cluster (Tinsley & Eldredge, 1995). If clusters or categories of activities are going to be used in future studies to analyze the relationship between leisure activities and life satisfaction, a classification system that takes into psychological needs may be more useful. For instance, Tinsley and Eldredge (1995) proposed using 12 clusters based on the psychological needs fulfilled by 82 general activities. These 12 clusters included agency, novelty, belongingness, service, sensual enjoyment, cognitive stimulation, self-expression, creativity, competition, vicarious competition, relaxation, and residual (cluster containing miscellaneous activities) (Tinsley & Eldredge, 1995).

The third reason for why leisure activities generally explain little to no variance in life satisfaction is that leisure activity participation is often analyzed as if it has a direct relationship to life satisfaction. As indicated above, studies have shown a weak relationship between leisure activity participation and life satisfaction. It is more likely that the relationship between leisure activities and life satisfaction is mediated by other factors. These may include leisure satisfaction (Lloyd & Auld, 2002; Ragheb, 1989), physical and mental health (Iso-Ahola, 1997), stress reduction (Iwasaki, Mannell, Smale, & Butcher, 2005), psychological needs (Tinsley & Tinsley, 1986), leisure coping strategies (Iwasaki, 2003), and positive and negative affect (Schimmack, 2008).

Leisure as an Experience

The meaning of leisure as a subjective experience has been extensively analyzed for over 40 years. Most of these efforts have focused on identifying key components of what constitutes a leisure experience for the largest amount of individuals. As an experience, leisure means "activity desires for its own sake (intrinsic desire)" (Cooper, 1999, p. 13); full autonomy, freedom, and control when participating in a leisure activity (Iso-Ahola, 1999); actions that are intrinsically motivated which can lead to feelings of enjoyment and a sense of freedom (Godbey, Caldwell, Floyd, & Payne, 2005). In other words, leisure refers to participating in leisure activities for intrinsically motivated or autonomous reasons. This means that a person must have "no internal or external pressures or coercion to engage in leisure activities. One participates in an activity because he or she finds it intrinsically interesting, for its own sake, out of sheer pleasure and enjoyment" (Iso-Ahola, 1999, p. 39). It is important to note that leisure activities themselves are not intrinsically or extrinsically interesting. "It is the individual who through psychological needs and processes finds some activities intrinsically and others extrinsically motivating" (Iso-Ahola, 1999, p. 39).

There are two other elements for participation in a leisure activity to be identified as a leisure experience, but these are not as agreed upon as the first element discussed. Like the component of intrinsic motivation, each element serves as motivations for and as benefits or positive outcomes of leisure activity participation.

Thus, in addition to intrinsic motivation, some sense of intrinsic reward should be obtained from participation in the leisure activity. An important reward, which some argue as the most important reward, is that of competence. Increased competence allows us to become more engaged in different activities such as sports, outdoor recreation, or even doing cross-word puzzles (Iso-Ahola, 1999).

The last element for what constitutes a leisure experience is whether it is a social experience or not. This "involves interrelationships, companionships, friendships, interaction and social support" (Iso-Ahola, 1999, p. 39). As a social experience, leisure provides a context for expression of social identity along with the development of intimate relationships.

It should be noted that there are other indicators of a leisure experience, but I have limited my discussion to the elements mentioned in this section. For example, Henderson (1990) identified components of leisure experiences for women as "including free choice or perceived freedom from constraints, intrinsic motivation, enjoyment, relaxation, role interactions, personal involvement, and self-expression" (p. 232). Similarly, in a sample of adolescents, components of leisure experiences were identified as participation in activities which exhibit a number of components including relaxation, enjoyment, intrinsic motivation, and happiness (Kleiber, Caldwell, & Shaw, 1993). Additionally, in a sample of adults, three indicators of a leisure experience emerged: intrinsic satisfaction (i.e., autonomy), perceived freedom, and involvement (Unger & Kernan, 1983).

Measures of Leisure Experience

Given the importance of the leisure experience to leisure researchers, it has been measured using a wide array of methods and methodologies. It has been measured using both qualitative and quantitative measures as well as ESM and other similar techniques such as the self-initiated-tape-recording method (SITRM). It has therefore been measured during activity participation real time as well as post-activity (Lee, Dattilo, & Howard, 1994).

Quality-of-Life Indexes for Leisure Experience

While leisure as free time and leisure as an activity are more commonly measured in quality-of-life indexes, at the present moment, I am not aware of any quality-of-life index which considers elements of leisure experience. While research has not guided us as to why this is so, one reason may be with the usability of leisure experience measures which commonly use multiple item indicators for multiple aspects of the leisure experience. Additionally, one shot measures often do not provide a clear understanding of leisure experiences when compared to other constructs, such as life satisfaction. I believe these issues have made it more challenging to include measures of leisure experience in modern quality-of-life indexes.

Leisure Experience and Life Satisfaction

In spite of the years spent on research focusing on leisure experience, there is surprisingly little research available looking at the relationship between leisure experience and life satisfaction. The primary reason for this is simply a lack of consensus among leisure scholars as to what is meant by leisure experience. In fact, the research that is available only indirectly relates leisure experience to life satisfaction.

For example, if we look at autonomy and intrinsic motivation as the key components of what makes participation in a leisure activity a leisure experience, then we may begin to hypothesize the relationship between leisure experience and life satisfaction. More specifically, when compared to other psychological needs (e.g., self-esteem, security, and money–luxury), autonomy has been identified as one of the most significant psychological needs (Ryan & Deci, 2000; Ryan, 1995; Sheldon, Elliot, Kim, & Kasser, 2001). For example, individuals who have high autonomy have been found to also be high in positive affect and self-esteem and less likely to experience negative emotions such as shame or guilt or experience boredom (Deci & Ryan, 1987, 1995; Ryan, 1995). In a study by MacLeod and Conway (2005), positive affect of individuals ($n = 84$) in a small community in London was found to be significantly positively correlated ($r = 0.40$, $p < 0.001$) with life satisfaction. Moreover, negative affect in the same sample was significantly negatively correlated ($r = -0.31$, $p < 0.01$) with life satisfaction. Similarly, Sheldon, Ryan, Deci and Kasser (2004) found that in a sample of 221 undergraduates positive affect positively correlated with life satisfaction ($r = 0.61$) and negative affect negatively correlated with life satisfaction ($r = -0.38$). Moreover, there was a significant main effect for relative autonomy on expected happiness (i.e., life satisfaction) ($\beta = 0.50$, $p < 0.01$) in a different sample of undergraduates ($n = 714$) (Sheldon et al., 2004). The findings that high positive affect and low negative affect are strong predictors of life satisfaction are consistent with the other literature (Schimmack, Radhakrishnan, Oishi, & Dzokoto, 2002; Suh, Diener, Oishi, & Triandis, 1998).

Time, Activity, or Experience?

If the purpose is to better understand the relationship between leisure and life satisfaction, which construct should one choose? The answer is that each meaning of leisure can potentially contribute to our understanding of the relationship between leisure and life satisfaction, but more research is needed with each meaning of leisure. For instance, having little to no leisure time can be detrimental to a person who literally works all of the time. We are humans and have limits before we break mentally and physically. However, in most developed countries at the moment, there seems to be ample leisure time to participate in leisure activities which can fulfill our various needs. Thus, measurements from a more psychological perspective, such as found in the experience sampling methodology (ESM) or day reconstruction method (DRM), may provide a better framework from which to understand the relationship between leisure time and life satisfaction. Regarding leisure activities, I have made a case against testing extent of leisure activity participation as a direct effect of life satisfaction. Instead, better understanding the mediating relationship of other variables between leisure activity and life satisfaction may help us to better understand the true relationship between leisure activities and life satisfaction. Additionally, I have pointed out some of the limitations of simply focusing on extent of leisure activity participation. Future research should consider using additional or alternative measures to supplement limitations found in these measures. Finally, from a leisure experience perspective, focusing on the component of autonomy in leisure experiences may help to better understand the complex nature of leisure experiences and life satisfaction.

Conclusion

As described by Cummins (1997), leisure "... is a slippery concept. Unlike the other domains where generally 'more is better', this does not apply to leisure. And as soon as some qualifier is introduced, such as 'quality leisure', it immediately overlaps with other domains" (p. 36). Cummins' (1997) comments are accurate. Leisure has been a slippery concept throughout the quality of life and subjective well-being literature. This slipperiness has caused much research to use measurement techniques that will not sufficiently help us to understand the specific meaning of leisure we are attempting to understand, and this same slipperiness has caused us to test relationships between leisure activity participation and life satisfaction in which we actually should have believed we were trying to retain the null hypothesis.

In this chapter, three distinct meanings of leisure were explored: leisure as time (leisure time), as an activity (leisure activity), and as an experience (leisure experience). This included their conceptualization, operationalization, and relationship to life satisfaction or terms that have been used synonymously with life satisfaction such as happiness or which are related to life satisfaction such as subjective well-being.

In Cummins' (1997) quote, he indicated how leisure often overlaps with other domains. According to Kelly and Kelly (1994), "Leisure is related to work, family, education, personal development, sexuality, and almost everything else rather than being a clearly distinct aspect of life" (p. 250). Again, Cummins' (1997) statements are accurate. Regardless if we look at leisure from the perspective of time, activity, or experience, it will always overlap with other domains of life. The reason is because much of what is important in our lives happens outside of work. We play with our children in the park, dance with our spouses at home, have dinner with our aging parents, and go out to the movies with friends all during our leisure time. If we are lucky, the activities we participate in take on a whole new meaning and we find ourselves wanting to participate in the activity regardless of what anyone else thinks we should do. That is the leisure experience.

References

Aguiar, M., & Hurst, E. (2007). Measuring trends in leisure: The allocation of time over five decades. *The Quarterly Journal of Economics, 122*(3), 969–1006.

Baker, D. A., & Palmer, R. J. (2006). Examining the effects of perceptions of community and recreation participation on quality of life. *Social Indicators Research, 75*, 395–418.

Bandura, A. (1977). Self-efficacy: Toward a unifying theory of behavioral change. *Psychological Review, 84*, 191–215.

Bittman, M., & Wajcman, J. (2000). The rush hour: The character of leisure time and gender equality. *Social Forces, 79*(1), 165–189.

Boelhouwer, J., & Stoop, I. (1999). Measuring well-being in the Netherlands: The SCP index from 1974 to 1997. *Social Indicators Research, 48*(1), 51–75.

Clemente, F., & Sauer, W. J. (1976). Life satisfaction in the United States. *Social Forces, 54*(3), 621–631.

Cooper, W. (1999). Some philosophical aspects of leisure theory. In E. L. Jackson, & T. L. Burton (Eds.), *Leisure studies: Prospects for the twenty-first century* (pp. 3–15). State College, PA: Venture Publishing, Inc.

Csikszentmihalyi, M., & Hunter, J. P. (2003). Happiness in everyday life: The uses of experience sampling. *Journal of Happiness Studies, 4*, 185–199.

Cummins, R. A. (1997). *Comprehensive quality of life scale: Adult*. Burwood, Melbourne: School of Psychology, Deakin University.

Deci, E. L., & Ryan, R. M. (1987). The support of autonomy and the control of behavior. *Journal of Personality and Social Psychology, 53*(6), 1024–1037.

Deci, E. L., & Ryan, R. M. (1995). Human autonomy as the basis for true self-esteem. In M. H. Kernis (Ed.), *Efficiency, agency, and self-esteem* (pp. 31–47). New York: Plenum Press.

Diener, E., Diener, M., & Diener, C. (1995). Factors predicting the subjective well-being of nations. *Journal of Personality and Social Psychology, 69*(5), 851–864.

Diener, E., Oishi, S., & Lucas, R. E. (2003). Personality, culture, and subjective well-being: Emotional and cognitive evaluations of life. *Annual Review of Psychology, 54*, 403–425.

Diener, E., & Seligman, M. E. P. (2002). Very happy people. *Psychological Science, 13*(1), 81–84.

Diener, E., Suh, E. M., Lucas, R. E., & Smith, H. L. (1999). Subjective well-being: Three decades of progress. *Psychological Bulletin, 125*(2), 276–302.

Godbey, G., Caldwell, L. L., Floyd, M., & Payne, L. (2005). Contributions of leisure studies and recreation and park management research to the active living agenda. *American Journal of Preventive Medicine, 28*, 150–158.

Hagerty, M. R., Cummins, R. A., Ferris, A. L., Land, K. C., Michalos, A. C., Peterson, M., et al. (2001). Quality of life indexes for national policy: Review and agenda for research. *Social Indicators Research, 55*, 1–96.

Henderson, K. A. (1990). The meaning of leisure for women: An integrative review of the research. *Journal of Leisure Research, 22*(3), 228–243.

Hills, P., & Argyle, M. (1998). Positive moods derived from leisure and their relationship to happiness and personality. *Personality and Individual Differences, 25*(3), 523–535.

Iso-Ahola, S. E. (1997). A psychological analysis of leisure and health. In J. T. Haworth (Ed.), *Work, leisure and well-being* (pp. 131–144). London: Routledge.

Iso-Ahola, S. E. (1999). Motivational foundations of leisure. In E. L. Jackson & T. L. Burton (Eds.), *Leisure studies: Prospects for the twenty-first century* (pp. 35–51). State College, PA: Venture Publishing, Inc.

Iwasaki, Y. (2003). Examining rival models of leisure coping mechanisms. *Leisure Sciences, 25*, 183–206.

Iwasaki, Y., Mannell, R. C., Smale, B. J. A., & Butcher, J. (2005). Contributions of leisure participation in predicting stress coping and health among police and emergency response services workers. *Journal of Health Psychology, 10*(1), 79–99.

Kahneman, D., Krueger, A. B., Schkade, D. A., Schwarz, N., & Stone, A. A. (2004a). The Day Reconstruction Method (DRM): Instrument documentation. Retrieved November 27, 2009, from http://sitemaker.umich.edu/norbert.schwarz/files/drm_documentation_july_2004.pdf

Kahneman, D., Krueger, A. B., Schkade, D. A., Schwarz, N., & Stone, A. A. (2004b). A survey method for characterizing daily life experience: The day reconstruction method. *Science, 306,* 1776–1780.

Kelly, J. R., & Kelly, J. R. (1994). Multiple dimensions of meaning in the domains of work, family, and leisure. *Journal of Leisure Research, 26*(3), 250–274.

Kleiber, D. A., Caldwell, L. L., & Shaw, S. M. (1993). Leisure meanings in adolescence. *Society and Leisure, 16*(1), 99–114.

Lane, P. M., & Lindquist, J. D. (1988). Definitions for the fourth dimension: A proposed time classification system. In K. D. Bahn (Ed.), *Developments in marketing science* (Vol. XI, pp. 38–46). Blacksburg, VA: Academy of Marketing Science.

Lee, Y., Dattilo, J., & Howard, D. (1994). The complex and dynamic nature of leisure experience. *Journal of Leisure Research, 26*(3), 195–211.

Lemon, B. W., Bengtson, V. L., & Peterson, J. A. (1972). An exploration of the activity theory of aging: Activity types and life satisfaction among in-movers to a retirement community. *Journal of Gerontology, 27*(4), 511–523.

Lewis, V. G., & Borders, L. D. (1995). Life satisfaction of single middle-aged professional women. *Journal of Counseling and Development, 74,* 94–100.

Lloyd, K. M., & Auld, C. J. (2002). The role of leisure in determining quality of life: Issues of content and measurement. *Social Indicators Research, 57,* 43–71.

MacLeod, A. K., & Conway, C. (2005). Well-being and the anticipation of future positive experiences: The role of income, social networks, and planning ability. *Cognition and Emotion, 19*(3), 357–374.

Mannell, R. C., & Kleiber, D. A. (1997). *A social psychology of leisure.* State College, PA: Venture Publishing, Inc.

Michalos, A. C. (2003). *Essays on the quality of life* (Vol. 19). Boston: Kluwer Academic Publishers.

Page-Wood, E. S., & Lane, P. M. (1989). *The leisure component in quality of life: Quality leisure and the fourth dimension.* Paper presented at the Third Quality-of-Life/Marketing Conference, Blacksburg, VA.

Ragheb, M. G. (1989). Step-wise regression analysis of leisure domains and the reported contribution of leisure activities to individuals' well-being: An exploratory study. *Society and Leisure, 12*(2), 399–412.

Ragheb, M. G. (1996). The search for meaning in leisure pursuits: Review, conceptualization and a need for a psychometric development. *Leisure Sciences, 15,* 245–258.

Ragheb, M. G., & Griffith, C. A. (1982). The contributions of leisure participation and leisure satisfaction to life satisfaction of older persons. *Journal of Leisure Research, 14*(4), 295–306.

Ray, R. O. (1979). Life satisfaction and activity involvement: Implications for leisure service. *Journal of Leisure Research, 11*(2), 112–119.

Riddick, C. C. (1985). Life satisfaction determinants of older males and females. *Leisure Sciences, 7*(1), 47–63.

Rodríguez, A., Látková, P., & Sun, Y. -Y. (2008). The relationship between leisure and life satisfaction: Application of activity and need theory. *Social Indicators Research, 86,* 163–175.

Russell, R. V. (2009). *Pastimes: The context of contemporary leisure* (4th ed). Champaign, IL: Sagamore Publishing.

Ryan, R. M. (1995). Psychological needs and the facilitation of integrative processes. *Journal of Personality, 63*(3), 397–427.

Ryan, R. M., & Deci, E. L. (2000). Self-determination theory and the facilitation of intrinsic motivation, social development, and well-being. *American Psychologist, 55*(1), 68–78.

Schimmack, U. (2008). The structure of subjective well-being. In M. Eid & R. J. Larson (Eds.), *The science of subjective well-being* (pp. 97–123). New York: Guilford Press.

Schimmack, U., Radhakrishnan, P., Oishi, S., & Dzokoto, V. (2002). Culture, personality, and subjective well-being: Integrating process models of life satisfaction. *Journal of Personality and Social Psychology, 82*(4), 582–593.

Schnohr, P., Kristensen, T. S., Prescott, E., & Scharling, H. (2005). Stress and life dissatisfaction are inversely associated with jogging and other types of physical activity in leisure time: The Copenhagen city heart study. *Scandinavian Journal of Medicine and Science in Sports, 15*, 107–112.

Scollon, C. N., Kim-Prieto, C., & Diener, E. (2003). Experience sampling: Promises and pitfalls, strengths and weaknesses. *Journal of Happiness Studies, 4*, 5–34.

Sheldon, K. M., Elliot, A. J., Kim, Y., & Kasser, T. (2001). What is satisfying about satisfying events? Testing 10 candidate psychological needs. *Journal of Personality and Social Psychology, 80*(2), 325–339.

Sheldon, K. M., Ryan, R. M., Deci, E. L., & Kasser, T. (2004). The independent effects of goal contents and motives on well-being: It's both what you pursue and why you pursue it. *Personality and Social Psychology Bulletin, 30*(4), 475–486.

Shin, D. C., & Rutkowski, C. P. (2003). Subjective quality of Korean life in 1981 and 2001. *Social Indicators Research, 62*(63), 509–534.

Silverstein, M., & Parker, M. G. (2002). Leisure activities and quality of life among the oldest old in Sweden. *Research on Aging, 24*(5), 528–547.

Sneegas, J. J. (1986). Components of life satisfaction in middle and later life adults: Perceived social competence, leisure participation, and leisure satisfaction. *Journal of Leisure Research, 18*(4), 248–258.

Suh, E. M., Diener, E., Oishi, S., & Triandis, H. C. (1998). The shifting basis of life satisfaction judgments across cultures: Emotions versus norms. *Journal of Personality and Social Psychology, 74*(2), 482–493.

Tinsley, H. E. A., & Eldredge, B. D. (1995). Psychological benefits of leisure participation: A taxonomy of leisure activities based on their need-gratifying properties. *Journal of Counseling Psychology, 42*(2), 123–132.

Tinsley, H. E. A., & Tinsley, D. J. (1986). A theory of the attributes, benefits, and causes of leisure experience. *Leisure Sciences, 8*(1), 1–45.

Unger, L. S., & Kernan, J. B. (1983). On the meaning of leisure: An investigation of some determinants of the subjective experience. *The Journal of Consumer Research, 9*(4), 381–392.

World Health Organization (2000a). WHOQoL–100: Australian Version. Retrieved November 29, 2009, from http://www.psychiatry.unimelb.edu.au/qol/whoqol/instruments/WHOQoL-100_instrument.pdf

World Health Organization (2000b). WHOQoL-BREF: Australian Version. Retrieved November 29, 2009, from http://cvwww.psychiatry.unimelb.edu.au/qol/whoqol/instruments/WHOQoL-BREF_instrument.pdf

Chapter 4
Publicly Accessible Space and Quality of Life: A Tool for Measuring the Openness of Urban Spaces

Jeremy Németh and Stephan Schmidt

Introduction

Safety and security are essential components of urban public space management, particularly since September 11, 2001. Although security is necessary for creating and maintaining publicly accessible spaces, making it a top priority is criticized as restricting social interaction, constraining individual liberties, and unjustly excluding certain populations. We argue that a focus on security and control over broader social goals such as openness and liberty can reduce the quality of life for particular individuals and groups. Therefore, this study examines legal, design, and policy tools used to exert social and behavioral control in publicly accessible urban spaces. Based on a review of the relevant literature as well as extensive site visits to public spaces in New York City, we create a comprehensive index that uses 20 separate indicators in four different broad categories to quantify the degree to which the use of a space is controlled. We demonstrate how the tool can be used and summarize the results of several recent applications. We then suggest several potential applications useful in planning practice and for testing theories about public space.

Vibrant public spaces are an integral part of the urban physical fabric, connecting disparate neighborhoods and encouraging interaction among an otherwise disparate constituency. At their best, public spaces can instill a sense of civic pride, encourage interaction among strangers, and promote inclusive democratic ideals. As such, the quality of public space, and of public life more generally, is directly dependent on its accessibility to a diversity of users. A more holistic conception of accessibility that moves beyond traditional interpretations of access as merely physical or visual is the concept of *openness*, defined herein as "the freedom or ability of people to achieve their basic needs in order to sustain their quality of life" (Lau & Chiu, 2003,

J. Németh (✉)
University of Colorado, Denver, CO, USA
e-mail: jeremy.nemeth@colorado.edu

This chapter is excerpted with permission from the article, 'Toward a Methodology for Measuring the Security of Publicly Accessible Spaces' Jeremy Németh and Stephen Schmidt *Journal of the American Planning Association*, *73*(3), 283–279. See http://www.informaworld.com.

p. 197). Thus a space which encourages freedom of use, behavior, and access is a more open and democratic space. Relating openness to quality of life requires us to more closely examine the factors that might limit or discourage freedom of use, such as the introduction of security measures and other actions taken to ostensibly protect users of public space.

In recent years, urban planners, geographers, and legal theorists have paid significant attention to security in public spaces as urban revitalization efforts are often fixated on the creation of safe spaces and the provision of public space is increasingly undertaken by the private sector (Davis, 1992; Ellin, 1996; Fyfe, 1998; Loukaitou-Sideris, 1996; Low, 2003; Pain, 2001). This emphasis on security has only been exacerbated since September 11, 2001 as owners and managers of parks and plazas frequently cite concerns over potential terrorist attacks as justification to increase behavioral control (Davis, 2001; Marcuse, 2002; Mitchell & Staeheli, 2005; Warren, 2002). However, some note that security concerns are nothing new, arguing that "the terrorist attacks ... did not so much launch a new debate about public space as serve to intensify one that already exists" (Mitchell, 2003, p. 4).

These security measures have been criticized for restricting social interaction, constraining individual liberties, "militarizing" space, and excluding certain populations through interrelated legal, design, and policy tools (Davis, 1992; Graham & Marvin, 2001; Kohn, 2004). Some have argued that making security a top priority reduces the quality of life for marginalized populations while engendering a wholesale retreat from social life and an "end of public culture" (Banerjee, 2001; Mitchell, 1995; Sennett, 1978; Sorkin, 1992).

Yet, few studies have empirically tested such assertions or documented actual methods and approaches used to secure such spaces. Scholars have failed to heed appeals such as William Whyte's (1988) call for a "stiff clarifying test" to assess public access rights. Mark Francis's (1989) claim that "the effect of control on public environments raises several issues in need of further empirical study and design" and his request for "... a study of the role of control in the design, management, and use of different public-space types" have not been adequately addressed (p. 168). Studies involving observation-based research have been limited. Most either analyze a singular approach to controlling space, such as the use of legal measures or design techniques, or fail to objectively assess control, operating instead from the situated, experiential points of view of young people (Katz, 1998, 2006; Valentine, 1996, 2004), women (Day, 1999; Pain, 2001; Ruddick, 1996), racial and ethnic minorities (Jackson, 1998; McCann, 2000), or homeless persons (Mitchell, 1995, 2003).

One reason for this lack of pragmatic research is the absence of an adequate tool with which to conduct such an analysis. We address this oversight by operationalizing and testing a comprehensive, conceptually grounded index to allow researchers, city officials, and concerned citizens to empirically quantify the degree to which behavioral control is exerted over users of publicly accessible spaces (Németh and Schmidt, 2007) We rely on relevant literature and empirical observations of spaces in New York City to create this index. This chapter answers recent calls by critical

scholars to bridge the gap between the theoretical understandings of social and political space and the actual lived experiences of physical and material spaces (Low &
Smith, 2006).

It is important to note that the term *public space* can be applied to a wide variety of social environments, from urban streets and sidewalks to suburban shopping
malls and movie theaters to the public forums and chat rooms of the Internet. These
locations vary along a continuum of relative publicness and can be categorized
according to concepts of ownership, management, and accessibility. In this chapter we focus exclusively on parks, squares, and plazas (both publicly and privately
owned) and refer to such sites as *publicly accessible spaces.*

Securing Space

There is a general consensus that "perceptions and feelings of personal safety are
prerequisites for a vital and viable city" (Oc & Tiesdell, 1999, p. 265). This argu-
ment continues that the key to creating safer areas is the peopling of publicly
accessible space, as the presence of others reassures users that there are an adequate
number of "eyes on the street" to deter criminals and maintain a safe environment
(Jacobs, 1961). This approach is based on two contentions: personal crime is more
likely to occur in bleak, deserted areas and *fear* of public space often stems from
the fact that there are very few people around. In his landmark study of New York
plazas, Whyte (1988) also demonstrated that use begets more use. In other words,
passersby are more likely to enter a heavily used space, and the busier a space is the
more users it will attract (until some critical mass is reached). This relationship is
self-reinforcing: In order for spaces to be perceived as safe they must be well used,
but those with a choice will only use spaces that they perceive as safe.

However, scholars criticize this emphasis on security on two major grounds.
First, the desire to attract a more orderly citizenry often comes at the expense
of certain individuals deemed objectionable or disorderly. As publicly accessible
spaces are increasingly organized around consumption, those who contribute to
the accumulation of capital by purchasing goods and services are welcomed, while
those who fail to contribute are discouraged (Turner, 2002, p. 543; see also Fyfe &
Bannister, 1998; Judd & Fainstein, 1999; Mitchell, 1995; Németh, 2006; Németh &
Schmidt, 2009; Schmidt, 2004; Zukin, 1995). Put differently, "purifying and pri-
vatizing spaces to enhance the consumption experience of some comes at a price
of social exclusion and a sense of increasing inequality for others" (Fyfe, 1998,
p. 7). As potential users or consumers might be turned away by unruly or unconven-
tional people, spaces must not be accessible to those "disorderly people [that] may
deter some citizens from gathering in the agora" (Ellickson, 1996, p. 1180). Critics
claim the exclusion of such undesirable individuals is often based on conceptions
of race, class, gender, or physical appearance (see Carmona, Heath, Oc, & Tiesdell,
2003; Cresswell, 1996; Flusty, 1994; Shields, 1989; Sibley, 1995). Some argue that
managers of publicly accessible spaces frequently fail to make the vital distinction
between identity and conduct:

The mere identity of a person as homeless [for example]... should never disqualify that person from using the space. On the other hand, if that person's conduct... becomes such a nuisance to others that they are fully prevented from enjoying that space, then that person may legitimately be asked to... leave the space (Kayden, New York City Department of City Planning & Municipal Art Society of New York, 2000, p. 147).

Second, the identification of undesirable people requires that users be segregated into categories using concepts of appropriateness and orderliness. Wekerle and Whitzman (1995) suggest that "the paradox is that the law and order response kills the city it is purporting to save. It deepens the divisions and the fear of the 'other' which are among the most harmful effects of fear and crime" (p. 6). While policing, surveillance, and strict use regulations might stimulate perceptions of safety, they can also contribute to "accentuating fear by increasing paranoia and distrust among people" (Ellin, 1996, p. 153). Giving security priority over other spatial considerations forces owners and managers to act as "spatial police, regulators of bodies in space, deciding who can do what and be where, and even when" (Sandercock, 1998, p. 166; see Berman, 1986; Ruddick, 1996). This trend begins to restrict all users' civil liberties and quality of life, just as it erodes the public realm and reduces the potential for democratic expression (Crawford, 1992; Sennett, 1978; Young, 2000).

While some scholars identify high usage as an indicator of a successful space (see Carmona et al., 2003; Kayden et al., 2000), others argue that use itself should not be the only measure of success. An underutilized space, for example, may offer people a quiet, contemplative place to withdraw from the stresses of urban life (Loukaitou-Sideris and Banerjee, 1998, p. 302). In addition, "activity alone is not a good gauge of the public values attached to a space.... use of an office tower plaza may be the result of a lack of meaningful alternatives" (Francis, 1989, p. 155). The goal of public space should extend beyond increasing the number of people who enter it to providing space that hosts a diversity of uses and users (Loukaitou-Sideris and Banerjee, 1998).

Political theorist Iris Young offers a normative ideal of publicly accessible space to which we subscribe. She argues that successful spaces must be universally accessible and must contribute to democratic inclusion by encouraging interaction between acquaintances and strangers (Young, 2000; see also Kohn, 2004). Such ideal spaces serve as "the material location where social interactions and public activities of *all* members of the public occur" (Mitchell, 2003, p. 131, emphasis added). However, this vision is utopian, and the ideal of a universally inclusive and unmediated space can never be met (Mitchell, 2003). Public space is not homogeneous, and "the dimensions and extent of its publicness are highly differentiated from instance to instance" (Smith and Low, 2006, p. 3). In addition, *the public* is a contested term and is constantly challenged and reformulated.

Consequently, ideal publicly accessible spaces are those that encourage social interaction among the most diverse set of users possible. However, we do not claim that the most open or accessible spaces are always the most successful or that they *necessarily* increase the quality of life of all users. Instead, successful public spaces adeptly balance liberty with personal security: While a mother with small child might prefer a secure and controlled environment, a homeless person or group of teenagers might favor spaces lacking such mediation. An index measuring levels of spatial control would allow people to make and test assertions about a successful space based on their own set of ideals.

Approaches to Controlling Publicly Accessible Space

We can effectively group spatial management techniques into hard (or active) control and soft (or passive) control measures (Loukaitou-Sideris & Banerjee, 1998, pp. 183–185). Hard control involves the use of surveillance cameras, private security guards, and legal measures to bar certain activities like soliciting, smoking, loitering, or disorderly behavior. Soft control focuses on more symbolic techniques, such as access restriction during non-business hours, small-scale urban design measures (e.g., spikes on ledges), or the removal of public restrooms or food vendors that might attract undesirable users (see Whyte, 1988). Oc and Tiesdell (1999, 2000) further divided these groups into four major approaches: regulatory, fortress, panoptic, and animated. We utilize these categories as the point of departure for constructing our index.

Laws and Rules

Under the general category of hard control, legal and regulatory measures signal the appropriate use of a space and, consequently, what types of persons are allowed. In this sense, laws are important signifiers of a space's "social meaning" (Blomley, Delaney, & Ford, 2001, p. xix). Since publicly accessible spaces are increasingly owned and managed by the private sector, they are sometimes subject to the prescriptions of the property owner, and the rules governing these spaces are often more variable and inconsistent than those in publicly owned spaces. Rules can be "flexibly and differentially enforced in order to sustain an illusion of openness while maximizing management's control" (Kohn, 2004, p. 13). In many cities, planning codes specify that private owners can stipulate what they deem reasonable rules of conduct, and they are not subject to the same regulation or oversight as public owners (Kayden et al., 2000).

Surveillance and Policing

Another hard control technique is the use of surveillance cameras and security patrols. Urban planners and property managers often support the use of security cameras as a means to reduce criminal activity and alleviate fears of crime. In recent years, the prevalence of cameras in public locations has increased dramatically, even as research linking surveillance and decline in crime has been anything but conclusive. Indeed, most studies conclude that crime in the most scrutinized locations had simply been displaced to other areas of the city (Fyfe & Bannister, 1998, p. 262).

Electronic surveillance also stimulates concerns over privacy and civil liberties; some critics argue that managers use cameras to identify and exclude undesirable users based on appearance alone (Ellin, 1996; Koskela, 2000; Shields, 1989).

The use of security personnel to maintain order is another popular technique. Business improvement districts (BIDs) often hire private security guards to patrol neighborhood and commercial areas for signs of disorder "that drive shoppers, and eventually store owners and citizens, to the suburbs" (Siegel, 1992, p. 43; see MacDonald, 1996, 1998, 2002). BIDs are dependent on property owners to pay operating expenses and, consequently, adopt the priorities of their corporate clients rather than concerning themselves with social equity or quality of life issues (Christopherson, 1994; Katz, 2001; Zukin, 1995). This is generally true of the priorities and mandates of private security guards, which differ significantly from those of the traditional public police force. The primary concern of the private guard is to protect the property and interests of those paying his/her salary rather than the public interest (Oc & Tiesdell, 1999, p. 272).

While studies have shown that people often feel safer in the presence of security personnel (Day, 1999; Fyfe & Bannister, 1998), the overabundance of security often generates suspicion that a space is not safe enough to operate without such a significant police presence. Put another way, "the social perception of threat becomes a function of the security mobilization itself, not crime rates" (Davis, 1990, p. 224). Whyte (1988) and others have decried the use of such highly elaborate policing tactics, arguing instead that good places are fundamentally self-policing (p. 158). Managers of urban spaces are now increasingly likely to prefer more indirect, secondary surveillance provided by the janitors, maintenance staff, valets, receptionists, and doorpersons working in the space or its immediate vicinity (Németh, 2004). As Jacobs (1961) maintained, "no amount of police can enforce civilization where the normal, casual enforcement of it has broken down" (p. 41; see Oc & Tiesdell, 1997, 1999).

Design and Image

Design, an example of soft control, can be used both literally and symbolically to control behavior and use of publicly accessible space. Christopherson (1994) describes how in response to real or perceived threats to security, urban designers and architects can (often at the behest of property managers or owners) specify rigid, orderly arrangements to control activity. These decisions can either "reinforce or challenge existing patterns of inclusion or exclusion" (Kohn, 2004, p. 7), because they can dictate appropriate spatial use and render a space less inviting to those failing to use it in such a manner. If a designer specifies benches outfitted with metal crossbars to prohibit reclining homeless people, it becomes clear that decisions concerning physical design have sociocultural consequences (Oc & Tiesdell, 1997). In this sense, "urban design organizes bodies socially and spatially... [I]t can stage and frame those who inhabit its spaces" (Rendell, 1998, p. 84).

Critics like Mike Davis (1990) have maintained that designers use their power to fortify publicly accessible spaces, transforming them into defensive (or defensible) bastions (see Ellin, 1996; Flusty, 1994; Mitchell, 1995). Whyte (1988) said that this response is calculated: Property owners often worry that if a place is made too attractive it will attract the very undesirable people they were trying to keep out in the first place. Physical redesigns become extremely attractive options since "design mechanisms are more expedient than having to legislate civility in public spaces" (Loukaitou-Sideris & Banerjee, 1998, p. 163).

However, many owners and managers of publicly accessible spaces believe, following Jane Jacobs, that the more people present in a space, the safer it will be. For this reason, managers entice potential users through measures that improve a space's image and, subsequently, increase overall usage. Such techniques include the introduction of public restrooms, food vendors or kiosks, movable chairs, flexible seating, sculptures, and interactive art, as well as an increased attention to environmental factors such as sun, nighttime lighting, wind, shadows, and trees. In addition, owners often pay for such improvements through lucrative sponsorship deals with private corporations who finance the particular upgrade or addition (such as the HSBC Bank Reading Room or Evian-sponsored umbrellas in New York City's Bryant Park or the Trump Organization's Wollman Skating Rink in Central Park). Critics lament the overuse of visible advertising in public spaces, arguing that users prefer publicly accessible spaces as a retreat from the often unrelenting visual stimulation of billboards, signs, and posters that dominate urban environments (Loukaitou-Sideris & Banerjee, 1998).

Access and Territoriality

Access restrictions and territorial separation are another set of soft control techniques frequently used both to attract and to deter specific users (see Newman, 1972). This is accomplished by programming certain areas for restricted or conditional use, such as cafes or restaurants which require patrons to pay in order to enter an area or sit at tables. While the programming of activities like chess, bocce, or exercising dogs attracts certain users, this practice has a tendency to restrict large areas to single uses only, leaving others to crowd into the leftover space. In this regard, the division of territory can segregate users by determining who can and cannot enter and who belongs in a particular area and who does not (Oc & Tiesdell, 1999, p. 270).

Carr, Francis, Rivlin, and Stone (1992) divided access into three different forms: visual, physical, and symbolic (from Carmona et al., 2003). Visual access is the ability to look into a space. Whyte (1988) noted that "if people do not see a space, they will not use it" (p. 129). He astutely observed that a plaza's relationship to the street and sidewalk is vital in attracting users into a space. By viewing just enough of a space to notice who is using it, and in what manner, people can quickly assess whether they would feel comfortable once inside.

Physical access involves one's actual ability to enter a space. Physical access is denied if the manager of a (supposedly) publicly accessible space closes the space's gates or locks its doors when it is legally required to be open or keeps a space open only to employees working in the building to which it is attached. Kayden (2005) also describes how private owners of publicly accessible spaces can deny access to patrons by barricading the space behind plywood or closing a space for construction for many months without apparent end (p. 126).

Symbolic access concerns whether one feels welcome in a space. Passing through a constricted entry, gate, or door or even through a security checkpoint can make visitors feel uncomfortable. Owners and managers of spaces can accentuate such feelings by placing a physical barrier such as a dumpster or scaffolding at a space's most convenient or natural entry point (Kayden, 2005, p. 126). Loukaitou-Sideris and Banerjee (1998) described how many privately owned public spaces, especially those intended to project a certain corporate image, tend to be introverted and physically disconnected from the broader public realm. Designers achieve this disconnection by setting the space several steps above or below the public sidewalk: "[O]nce past three feet a space can become relatively inaccessible. . . it is not only a physical matter so much as a psychological one" (Whyte, 1988, p. 129).

An Index to Measure the Control of Publicly Accessible Spaces

These four major approaches form the basis of our index. In order to construct the index, we also relied on site visits to 171 publicly accessible spaces in New York City, which has over 1,700 parks, playgrounds, and recreation facilities totaling 28,000 acres (New York Parks and Recreation, 2006) and more than 500 privately owned but publicly accessible spaces comprising 82 aggregate acres (Kayden et al., 2000).

We limited our study to midtown Manhattan, a high-density pedestrian area with the greatest concentration of highly trafficked publicly accessible spaces in the city. Coterminous with the boundaries of Community Board 5 and roughly bounded on the south by 14th Street and on the north by 59th Street, the area includes many higher profile corporate headquarters and places of business whose owners and occupants view security as an important priority. We acknowledge that limiting the fieldwork to midtown Manhattan may have made our index less generalizable. Nevertheless, we feel these sites present a unique opportunity. First, the security of publicly accessible spaces, both public and private, is an especially significant issue in New York City, particularly since September 11, 2001. In addition, choosing heavily trafficked, high-profile spaces allows us to witness spatial control where it is most prominent and deliberate.

We visited publicly accessible spaces with different ownership and management regimes in order to sample as wide a variety of spaces as possible. These include parks and places that are privately owned and managed (e.g., Trump Tower), publicly owned but privately managed (e.g., Bryant Park), and publicly owned and managed (e.g., Union Square). While control measures in publicly owned and managed spaces must conform to generally uniform standards,[1] the New York City Department of Planning stipulates that privately owned and managed spaces must only provide "reasonable rules of conduct" similar to those that apply in publicly owned parks and plazas (Kayden et al., 2000, p. 39). In fact, rules and regulations in privately owned spaces need not be officially reviewed by the department (P. Schneider, New York Department of City Planning, personal communication, February 28, 2007). Although the right to free speech and assembly in publicly owned spaces is guaranteed by the First Amendment, this right does not necessarily extend to spaces owned and managed by the private sector.[2] For these reasons, privately owned or managed spaces often employ very different security and control measures from those used in publicly owned and managed spaces.

Based on our site visits, we operationally defined 20 variables for the index, grouped into the aforementioned categories, each of which represents a possible strategy for securing space (Oc & Tiesdell, 1999, p. 270). The 20 variables are divided into 10 indicating control of users and 10 indicating free use of the space. Table 4.1 describes the variables and scoring criteria for each, while Tables 4.2 and 4.3 provide a more detailed description of each variable.

Reliability

Since people's perceptions of space are both variable and subjective, we aimed to make the index more reliable by specifying objective, directly observable indicators, and provided a scoring rubric (0, 1, or 2) based on the presence and intensity of each variable. The overall index score for a given space should be calculated by subtracting the total score for all variables indicating control from the total score for all variables indicating free use. The lower the score, on both individual variables and overall, the more controlled the space, and the higher the score, the freer the use of the space. The highest possible overall score is 20 (least controlled), the lowest is −20 (very controlled); zero would indicate a perfectly neutral score. For illustrative purposes, the Appendix depicts the total scores for two publicly accessible spaces: Washington Square Park and Sony Plaza.

Since many of the better designed spaces in cities are of "a higher quality that now attracts the very public that some owners then attempt to discourage from using the space" (Kayden, 2005, p. 125), managers are often compelled to introduce additional security measures. Conversely, managers of underutilized public spaces may have no need to implement strong security regimes since there are no users to control (Lees, 1998). Therefore, the index accounts for variables that control users as well as variables that encourage free use to prevent better designed and more used spaces from scoring lower on the index (appearing more controlling) than their underutilized counterparts.

Validity

The index was then validated by a panel of experts, including two academics in planning and design and two practitioners in the field of urban planning and landscape architecture. The panelists were selected based on their knowledge of

Table 4.1 Index variables

	Approach	Scoring criteria
Features that control users		
Visible sets of rules posted	Laws/rules	0 = none present 1 = one sign or posting 2 = two or more signs
Subjective/judgment rules posted	Laws/rules	0 = none present 1 = one rule visibly posted 2 = two or more rules visibly posted
In business improvement district (BID)	Surveillance/policing	0 = not in a BID 1 = in a BID with maintenance duties only 2 = in a BID with maintenance and security duties
Security cameras	Surveillance/policing	0 = none present 1 = one stationary camera 2 = two or more stationary cameras or any panning/moving camera
Security personnel	Surveillance/policing	0 = none present 1 = one private security guard or up to two public security personnel 2 = two or more private security or more than two public personnel
Secondary security personnel	Surveillance/policing	0 = none present 1 = one person or space oriented toward reception 2 = two or more persons or one person w/ space oriented at reception
Design to imply appropriate use	Design/image	0 = none present 1 = only one or two major examples 2 = several examples throughout space
Presence of sponsor/advertisement	Design/image	0 = none present 1 = one medium sign or several small signs 2 = large sign or two or more signs

Table 4.1 (continued)

	Approach	Scoring criteria
Areas of restricted or conditional use	Access/territoriality	0 = none present 1 = one small area restricted to certain members of the public 2 = large area for consumers only or several smaller restricted areas
Constrained hours of operation	Access/territoriality	0 = open 24 h/day, 7 days/week, most days of year 1 = at least part of space open past business hours or on weekends 2 = open only during business hours or portions permanently closed
Features encouraging freedom of use Sign announcing "public space"	Laws/rules	0 = none present 1 = one small sign 2 = one large sign or two or more signs
Public ownership/management	Surveillance/policing	0 = privately owned and privately managed 1 = publicly owned and privately managed 2 = publicly owned and publicly managed
Restroom available	Design/image	0 = none present 1 = available for customers only or difficult to access 2 = readily available to all
Diversity of seating types	Design/image	0 = no seating 1 = only one type of stationary seating 2 = two or more types of seating or substantial movable seating
Various microclimates	Design/image	0 = no sun or no shade or fully exposed to wind 1 = some sun/shade, overhangs/shielding from wind and rain 2 = several distinct microclimates, extensive overhangs, trees
Lighting to encourage nighttime use	Design/image	0 = none present 1 = one type or style of lighting 2 = several lighting types (e.g., soft lighting, overhead, lampposts)

Table 4.1 (continued)

	Approach	Scoring criteria
Small-scale food consumption	Design/image	0 = none present 1 = one basic kiosk or stand 2 = two or more kiosks/stands or one larger take-out stand
Art/cultural/visual enhancement	Design/image	0 = none present 1 = one or two minor installations, statues, or fountains 2 = one major interactive installation or frequent free performances
Entrance accessibility	Access/territoriality	0 = gated or key access only, and at all times 1 = one constricted entry or several entries through doors/gates only 2 = more than one entrance without gates
Orientation accessibility	Access/territoriality	0 = not on street level or blocked off from public sidewalk 1 = street level but oriented away from public sidewalk 2 = visible with access off sidewalk (or fewer than five steps up/down)

Table 4.2 Variable definitions: features that control users

Laws/rules	
Visible sets of rules posted	Official, visible signs listing *sets* of rules and regulations (not individual rules) on permanent plaques or "table tents." Listed rules should generally be objective and easily enforceable, like prohibitions against smoking, sitting on ledges, passing out flyers without permit, or drinking alcohol
Subjective/judgment rules posted	Official, visible signs listing individual rules describing activities prohibited after personal evaluations and judgments of desirability by owners, managers, or security guards. Such rules include no disorderly behavior, no disturbing other users, no loitering, no oversized baggage, or appropriate attire required
Surveillance/policing	
In business improvement district (BID)	Spaces located in business improvement districts (BIDs) are more likely to have electronic surveillance and private security guards and less likely to include public input into decisions regarding park management. BIDs can employ roving guards to patrol especially problematic neighborhood spaces
Security cameras	Although cameras must be visible to observer to be counted, many cameras are hidden from view. Cameras are often located inside buildings or on surrounding buildings but are oriented toward space. Stationary cameras are more common, often less intimidating than moving/panning cameras
Security personnel	Scoring dependent on time of visit. Publicly funded police, park rangers, private security guards. For index, score only when security is dedicated to space. Since private security only directed by property owner, often more controlling (and score higher on index) since police trained more uniformly
Secondary security personnel	Scoring dependent on time of visit. Includes maintenance staff, doorpersons, reception, cafe or restaurant employees, bathroom attendants. Also, spaces often oriented directly toward windowed reception or information area to ensure constant employee supervision

Table 4.2 (continued)

Design/image	
Design to imply appropriate use	Small-scale design to control user behavior or imply appropriate use. Examples might include metal spikes on ledges; walls, barriers, bollards to constrict circulation or to direct pedestrian flow; rolled, canted, or overly narrow and unsittable ledges; or crossbars on benches to deter reclining
Presence of sponsor/advertisement	Signs, symbols, banners, umbrellas, plaques tied to space's infrastructure, not to immediate services provided (e.g., cafés, kiosks). While non-advertised space is important for seeking diversion from city life, sponsored signs/plaques can push sponsors to dedicate resources for upkeep since company name is visible
Access/territoriality	
Areas of restricted/conditional use	Portions of space off-limits during certain times of day, days of week, or portions of year. Can also refer to seating/tables only open to cafe patrons, bars open only to adults, dog parks, playgrounds, corporate events open to shareholders only, spaces for employees of surrounding buildings only
Constrained hours of operation	While some spaces are permitted to close certain hours of day, spaces not open 24 h inherently restrict usage to particular population. Also, while usually due to lack of adequate supervision, spaces open only during weekday business hours clearly prioritize employee use over general public

Table 4.3 Variable definitions: features that encourage freedom of use

Laws/rules	
Sign announcing public space	Most zoning codes require publicly accessible spaces to exhibit plaques indicating such. Some spaces are clearly marked with signs denoting their public nature (e.g., New York's Sony Plaza), but when a sign or plaque is hidden by trees/shrubs or has graffiti covering it, its intent becomes null
Surveillance/policing	
Public ownership/management	Could fall in "laws/rules" approach, but more likely to impact type/amount of security, electronic surveillance in a space. Management often by conservancy or restoration corporation. Spaces can be publicly owned/publicly managed, publicly owned/privately managed, or privately owned/privately managed
Design/image	
Restroom available	Clearly some spaces are not large enough to merit public restroom. Realizing that free public restrooms often attract homeless persons, managers often remove them altogether or locate them in onsite cafes or galleries available to paying customers only (or providing keyed access for "desirable" patrons)
Diversity of seating types	Amount of seating is often most important factor for encouraging use of public space. Users often evaluate entry to space based on amount of available seating and ability to create varying "social distances." Movable chairs allow maximum flexibility and personal control in seating choice
Various microclimates	Spaces with various microclimate enclaves enlarge choice and personal control for users. Potential features might include shielding from wind; overhangs to protect from rain; areas receiving both sun and shade during day; or trees/shrubs/grass to provide connection with natural landscape
Lighting to encourage nighttime use	Studies indicate that vulnerable populations often avoid public spaces at night if not well lit. Lighting spaces encourages 24-h use, which has been shown to make visitors feel safer/more secure. However, critics argue that night lighting aids surveillance efforts and implies authoritative control

Table 4.3 (continued)

Small-scale food consumption	Most agree that food vendors enhance activity and vitality. This variable only includes small cafés, kiosks, carts, or stands selling food, drinks, or simple convenience items. Sit-down restaurants, clothing stores, and other full-scale retail establishments are not described by this variable
Art/cultural/visual enhancement	Art and aesthetic attraction can encourage use. Variable can include stationary visual enhancements like statues, fountains, or sculptures, also rotating art exhibits, public performances, farmers' markets, street fairs. Interactive features encourage use and personal control by curious patrons (often children)
Access/territoriality	
Entrance accessibility	If a space has locked doors or gates, requires a key to enter, or has only one constricted entry, it often feels more controlled or private than one with several non-gated entrances. In indoor spaces where users must enter through doors or past checkpoints, symbolic access and freedom of use diminished
Orientation accessibility	Spaces must be well integrated with sidewalk and street, as those oriented away from surrounding sidewalk, or located several feet above or below street level, make space less inviting. Well-used spaces are clearly visible from sidewalk, and users should be able to view surrounding public activity

New York City (where the assessed spaces were located) and their expertise in the design and management of public spaces. First, we scored eight popular spaces using the index. Each panelist was then shown several representative photographs of these spaces. Panelists were asked to assess each a subjective score (from –20 to +20) based on the perceived openness of the space. The criteria used for evaluation were more subjective and experiential than the 20 variables included in the index. A simple correlation between the panelists' subjectively assigned scores and the actual overall scores on the index revealed a statistically significant correlation coefficient (r^2) of 0.81 ($p < 0.05$). This suggests that the two measures are similar and the index captures the phenomenon of interest.

Testing the Index

To assess the feasibility of the index, we document results of three initial studies that all utilized the index. Each study served to further validate the index and demonstrate its reliability via its deployment by several scorers in different contexts (but all in New York City).

First, we pilot tested the index in the 12 busiest parks and plazas in central midtown Manhattan. We visited the five largest, busiest, and most complex privately owned and managed spaces in the area, those identified as "destinations" by Kayden et al. (2000).[3] We then chose four publicly owned and managed spaces and three publicly owned and privately managed spaces, each of which also exhibited the characteristics of a destination space.[4] In separate and independent visits conducted during busy weekday lunchtime hours (between 10 a.m. and 2 p.m.) we rated each space according to the index. Results from simple t tests determined that scores from our two separate sets of visits were not significantly different, providing some evidence that our scoring system was rigorous enough to be objectively employed.

Second, a published study by Németh (2009) applied the index to 163 privately owned and managed spaces in New York City. Overall scores for each space clustered relatively uniformly around the zero score. This application produced such an extensive number of data points that it allowed the author to conduct a principal components analysis on the spaces. He determined that seven basic approaches to spatial control exist, ranging from the sorting and filtering of users to the outright exclusion of visitors through strict access controls.

Third, a study by Németh and Schmidt (in press) examined a subset of 151 publicly accessible spaces to determine whether differences exist among spaces that are publicly owned and those that are privately owned and what the nature of those differences are. The sites included 89 privately owned and managed spaces and 62 publicly owned and managed spaces. The study had several statistically significant findings; first, as expected, privately owned public spaces tended to be more controlled or behaviorally restrictive than publicly owned spaces; second, while both publicly and privately owned spaces tend to equally encourage public use and access, privately owned spaces additionally include elements which tend to control use and behavior within those spaces, and specifically, privately owned spaces tended to employ additional surveillance/policing and access/territoriality restrictions than publicly owned spaces, although publicly owned spaces tended to have more rules and regulations.

A Demonstration of the Scoring Process

We demonstrate the scoring process with the following examples. Figures 4.1 and 4.2 depict the scoring for the variable *subjective or judgment rules posted*. Figure 4.1 shows that although the official rules of Washington Square Park urges users to "be courteous and respectful," none of the rules are necessarily subjective in nature. Based on the scoring criteria in Table 4.1, the space received 0 on this variable. Figure 4.2 shows a typical table tent located on each of Sony Plaza's 104 public tables. The rules prohibit "disorderly conduct," "obscene gestures," and "creating conditions that disturb others," regulations that can only be enforced after a judgment by the enforcer. We scored the space 2 on this variable.

Figures 4.3 and 4.4 demonstrate the scoring of the variable *restroom available*. Figure 4.3 shows the outdoor, visible, publicly accessible restroom in Washington

Fig. 4.1 Washington Square Park rules

Fig. 4.2   Sony Plaza rules

Fig. 4.3   Washington Square Park restroom

Square Park. Figure 4.4 depicts the restroom at Sony Plaza, which was difficult to access as it is located inside the Sony Wonder exhibit. Following our criteria, we scored Washington Square Park 2 and Sony Plaza 1, since entry to the restroom was monitored during the day by Sony employees.

Fig. 4.4 Sony Plaza restroom
location inside exhibit

As a final example, Figs. 4.5 and 4.6 illustrate how we scored the variable *entrance accessibility*. Figure 4.5 shows 1 of the 12 major entry points to Washington Square Park without gates and open 24 h. Figure 4.6 shows the entry to Sony Plaza through a set of glass doors. Because the entrant must cross through the symbolic barrier to the space, Sony Plaza received 1. Washington Square Park, with its multiple, ungated entry points, was scored 2 on this variable.

We encourage those using the index to obtain empirical data through independent visits by various users, owners, managers, or experts familiar with the spaces under examination. Separate observers should conduct multiple visits to the subject spaces to further check the validity of the index. This is particularly important when the index is applied outside New York City, since New York is an unusual case, as described above.

In addition, certain variables are time dependent. For example, a scorer may enter a space and find no security guards visible during the time he/she occupies the space. Another scorer may visit the space the following day or week and find several security guards present. In such cases, we recommend adopting the higher score, as it is obvious that the particular space does have several security personnel. To improve reliability, observers should make multiple visits, recording the time and date of each visit.

Fig. 4.5 Washington Square Park entry

Fig. 4.6 Sony Plaza entry

Conclusions and Future Research

This chapter proposes a comprehensive methodological tool to allow researchers, city officials, and citizens to empirically evaluate the degree to which control is exerted over users of publicly accessible spaces. However, we do not assert that a particular score on the index denotes a certain quality of life for a user. Quality of life is subjective and variable, as is one's desire to feel safe, secure, or controlled in public space. But we do argue that as behavioral control increases, individual choice decreases, and one's ability to position himself/herself in a situation that can *improve* our quality of life wanes. Therefore, the index is an important barometer that can be coupled with additional social indicators in measuring quality of life. We recommend social researchers add it to the burgeoning toolbox of indicators introduced in this volume and others.

As the index enables empirical testing of a number of questions, we suggest several potential applications for urban researchers, planners and policymakers, and neighborhood or community groups. First, are public spaces becoming increasingly restrictive and controlled over time, as some critics claim (Banerjee, 2001; Kohn, 2004)? The systematic application of our index allows researchers to monitor the changing presence and intensity of control, as owners and managers frequently update or alter their space's security measures. Second, Day (1999) and others argue that some of the most popular publicly accessible spaces are those that exert the most behavioral control over users. The index allows empirical testing of this and similar allegations correlating success, use, and control. Third, do certain socioeconomic or demographic populations prefer particular levels of spatial security? Researchers can apply the index to several sites and then compare scores with user counts and preference surveys to determine whether significant correlations exist.

Practitioners can apply these research findings to improve the design and maintenance of publicly accessible spaces and bring more balance to discussions concerning security and freedom of access and use. The index also enables planners, policymakers, and business improvement district (BID) officials to more efficiently and effectively assess levels of spatial control and adjust these levels based on a set of predetermined criteria. For example, planners could determine an ideal level of security based on the level of crime in a particular block or neighborhood. The application of the index could then suggest the need for an increase or decrease in the presence and intensity of certain measures. Similarly, planners are able to assess whether, as is commonly believed, crime rates and security levels are inversely related and determine which measures might be more or less effective in reducing criminal activity.

Finally, the index has important applications for neighborhood and community groups, local residents, students, and public and private organizations concerned about the steady erosion of civil liberties in the public realm. It can provide empirical evidence to defend claims that publicly accessible spaces are, in fact, becoming less accessible. Additionally, publicly available, interactive, and real-time scoring of such public spaces may be of interest to these parties. In this regard, the index can also serve to engage local communities and help facilitate public participation in the production of the built environment.

Notes

1. For list of official rules and regulations in New York City's parks, see New York City Department of Parks & Recreation's *Rules* (1990) or its Web site http://www.nycgovparks.org (2007). These sources note that standard rules apply to all publicly owned and managed spaces unless expressly noted in a site-specific ordinance or approved variance.
2. A number of Supreme Court decisions have addressed the rights to speech and assembly in privately owned spaces: *Marsh v. Alabama* (1946); *Amalgamated Food Employees Union v. Logan Valley Plaza* (1968); *Lloyd Corp. v. Tanner* (1972); *Hudgens v. National Labor Relations Board* (1976); and *Pruneyard Shopping Center v. Robins* (1980).
3. Destination space is broadly defined by Kayden et al. (2000) as "high-quality public space that attracts employees, residents, and visitors from outside, as well as from the space's immediate neighborhood" (p. 50).
4. We visited 12 spaces for the pilot test. Five were privately owned and managed: IBM Building, Rockefeller Center, Sony Plaza, Trump Tower, and World Wide Plaza. Three were publicly owned but privately managed: Bryant Park, Greeley Square, and Herald Square. Four were publicly owned and managed: Grand Army Plaza, Madison Square Park, Union Square Park, and Washington Square Park.

Appendix

Table 4.4 Sample scoring sheet comparing two random spaces: Washington Square Park and Sony Plaza

Features that control users	Approach	Wash. Sq. Park	Sony Plaza
Visible sets of rules posted	Laws/rules	2	2
Subjective/judgment rules posted	Laws/rules	0	2
In business improvement district (BID)	Surveillance/policing	0	2
Security cameras	Surveillance/policing	0	2
Security personnel	Surveillance/policing	1	2
Secondary security personnel	Surveillance/policing	0	2
Design to imply appropriate use	Design/image	0	1
Presence of sponsor/advertisement	Design/image	0	2
Areas of restricted or conditional use	Access/territoriality	2	2
Constrained hours of operation	Access/territoriality	1	1
Total		6	18
Features encouraging freedom of use			
Sign announcing "public space"	Laws/rules	2	2
Public ownership/management	Surveillance/policing	2	0
Restroom available	Design/image	2	1

Table 4.4 (continued)

Features that control users	Approach	Wash. Sq. Park	Sony Plaza
Diversity of seating types	Design/image	2	2
Various microclimates	Design/image	2	1
Lighting to encourage nighttime use	Design/image	2	1
Small-scale food consumption	Design/image	1	2
Art/cultural/visual enhancement	Design/image	2	1
Entrance accessibility	Access/territoriality	2	1
Orientation accessibility	Access/territoriality	2	2
Total		19	13
Overall score		13	−5

References

Banerjee, T. (2001). The future of public space: Beyond invented streets and reinvented places. *Journal of the American Planning Association, 67*(1), 9–24.

Berman, M. (1986). Take it to the streets: Conflict and community in public space. *Dissent, 33*(4), 476–485.

Blomley, N., Delaney, D., & Ford, R. (Eds.). (2001). *The legal geographies reader.* Oxford: Blackwell.

Carmona, M., Heath, T., Oc, T., & Tiesdell, S. (2003). *Public places – Urban spaces.* Oxford: Architectural Press.

Carr, S, Francis, M, Rivlin, L and Stone, A. (1992). Public space. Cambridge: Cambridge University Press.

Christopherson, S. (1994). The fortress city: Privatized spaces, consumer citizenship. In A. Amin (Ed.), *Post-fordism: A reader* (pp. 409–427). Oxford: Blackwell.

Crawford, M. (1992). The world in a shopping mall. In M. Sorkin (Ed.), *Variations on a theme park: The new American city and the end of public space* (pp. 181–204). New York: Hill and Wang.

Cresswell, T. (1996). *In place/out of place: Geography, ideology, and transgression.* Minneapolis, MN: University of Minnesota Press.

Davis, M. (1990). *City of quartz: Excavating the future in Los Angeles.* London: Verso.

Davis, M. (1992). Fortress Los Angeles: The militarization of urban space. In M. Sorkin (Ed.), *Variations on a theme park: The new American city and the end of public space* (pp. 154–180). New York: Hill and Wang.

Davis, M. (2001). The flames of New York. *New Left Review, 12,* 34–50.

Day, K. (1999). Introducing gender to the critique of privatized public space. *Journal of Urban Design, 4*(2), 155–178.

Ellickson, R. (1996). Controlling chronic misconduct in city spaces: Of panhandlers, skid rows, and public space zoning. *Yale Law Journal, 105*(5), 1165–1248.

Ellin, N. (1996). *Postmodern urbanism.* New York: Princeton Architectural Press.

Flusty, S. (1994). *Building paranoia: The proliferation of interdictory space and the erosion of spatial justice.* West Hollywood, CA: Los Angeles Forum for Architecture and Urban Design.

Francis, M. (1989). Control as a dimension of public space quality. In I. Altman and E. Zube (Eds.), *Public places and spaces: Human behavior and environment, Volume 10* (pp. 147–172). New York: Plenum.

Fyfe, N. (Ed.). (1998). *Images of the street: Planning, identity and control in public space*. New York: Routledge.

Fyfe, N., & Bannister, J. (1998). The "eyes upon the street": Closed-circuit television surveillance and the city. In N. Fyfe (Ed.), *Images of the street: Planning, identity and control in public space* (pp. 254–267). New York: Routledge.

Graham, S., & Marvin, S. (2001). *Splintering urbanism: Networked infrastructures, technological mobilities and the urban condition*. London: Routledge.

Jackson, P. (1998). Domesticating the street: The contested spaces of the high street and the mall. In N. Fyfe (Ed.), *Images of the street: Planning, identity and control in public space* (pp. 176–191). New York: Routledge.

Jacobs, J. (1961). *The death and life of great American cities*. New York: Vintage Books.

Judd, D., & Fainstein, S. (1999). The tourist city. New Haven, CT: Yale University Press.

Katz, C. (1998). Disintegrating developments: Global economic restructuring and the eroding of ecologies of youth. In T. Skelton and G. Valentine (Eds.), *Cool places: Geographies of youth culture* (pp. 130–144). London: Routledge.

Katz, C. (2001). Hiding the target: Social reproduction in the privatized urban environment. In C. Minca (Ed.), *Postmodern geography: Theory and praxis* (pp. 93–112). Oxford: Blackwell.

Katz, C. (2006). Power, space and terror: Social reproduction and the public environment. In S. Low and N. Smith (Eds.), *The politics of public space* (pp. 105–122). New York: Routledge.

Kayden, J., New York City Department of City Planning and the Municipal Art Society of New York (2000). Privately owned public space: The New York City experience. New York: Wiley.

Kayden, J. (2005). Using and misusing law to design the public realm. In E. Ben-Joseph and T. Szold (Eds.), *Regulating place: Standards and the shaping of urban America* (pp. 115–140). New York: Routledge.

Kohn, M. (2004). *Brave new neighborhoods: The privatization of public space*. New York: Routledge.

Koskela, H. (2000). The gaze without eyes: Video-surveillance and the changing nature of urban space. *Progress in Human Geography, 24*(2), 243–265.

Lau, J., & Chiu., C. (2003). Accessibility of low-income workers in Hong Kong. *Cities, 20*(3), 197–204.

Lees, L. (1998). Urban renaissance and the street: Spaces of control and contestation. In N. Fyfe (Ed.), *Images of the street: Planning, identity and control in public space* (pp. 236–254). New York: Routledge.

Loukaitou-Sideris, A. (1996). Cracks in the city: Addressing the constraints and potentials of urban design. *Journal of Urban Design, 1*(1), 91–104.

Loukaitou-Sideris, A., & Banerjee, T. (1998). *Urban design downtown: Poetics and politics of form*. Berkeley, CA: University of California Press.

Low, S. (2003). *Behind the gates: Life, security, and the pursuit of happiness in fortress America*. New York: Routledge.

Low, S. and Smith, N. (Eds.). (2006). *The politics of public space*. New York: Routledge.

MacDonald, H. (1996). BIDs really work. *City Journal*, (Spring).

MacDonald, H. (1998). BIDing adieu. *City Journal*, (Autumn).

MacDonald, H. (2002). Holiday homelessness hype. *City Journal*, (December).

Marcuse, P. (2002). Urban form and globalization after September 11th: The view from New York. *International Journal of Urban and Regional Research, 26*(3), 596–606.

McCann, E. (2000). Race, protest and public space: Contextualizing Lefebvre in the U.S. city. *Antipode, 31*(2), 163–184.

Mitchell, D. (1995). The end of public space? People's Park, definitions of the public, and democracy. *Annals of the Association of American Geographers, 85*, 108–133.

Mitchell, D. (2003). *The right to the city: Social justice and the fight for public space*. New York: The Guilford Press.

Mitchell, D., & Staeheli, L. (2005). Permitting protest: Parsing the fine geography of dissent in America. *International Journal of Urban and Regional Research, 29*(4), 796–813.

New York City Parks and Recreation (2006). *Frequently asked questions.* Retrieved November 1, 2006, from http://www.nycgovparks.org

Newman, O. (1972). *Defensible space: Crime prevention through urban design.* New York: Macmillan.

Németh, J. (2004). Redefining security in public space: The case of LOVE Park. *IEEE Technology and Society, 23*(4), 19–20.

Németh, J. (2006). Conflict, exclusion, relocation: Skateboarding and public space. *Journal of Urban Design, 11*(3), 297–318.

Németh, J. (2009). Defining a public: The management of privately owned public space. *Urban Studies, 46*(11), 1–28.

Németh, J., & Schmidt., S. (2007). Toward a methodology for measuring the security of publicly accessible spaces. *Journal of the American Planning Association, 73*(3), 283–297.

Németh, J. and Schmidt S. (2010). Space, place and the city: Emerging research on public space design and planning. Journal of Urban Design, 15(4):453–457.

Oc, T. and Tiesdell, S. (Eds.). (1997). *Safer city centres: Reviving the public realm.* London: Paul Chapman Publishing Ltd.

Oc, T., & Tiesdell, S. (1999). The fortress, the panoptic, the regulatory and the animated: Planning and urban design approaches to safer city centers. *Landscape Research, 24*(3), 265–286.

Oc, T., & Tiesdell, S. (2000). Urban design approaches to safer city centers: The fortress, the panoptic, the regulatory and the animated. In J. Gold and G. Revill (Eds.), *Landscapes of defence* (pp. 188–208). London: Prentice Hall.

Pain, R. (2001). Gender, race, age and fear in the city. *Urban Studies, 38*(5–6), 899–913.

Rendell, J. (1998). Displaying sexuality: Gendered identities and the early nineteenth-century street. In N. Fyfe (Ed.), *Images of the street: Planning, identity and control in public space* (pp. 75–91). New York: Routledge.

Ruddick, S. (1996). Constructing difference in public spaces: Race, class and gender as interlocking systems. *Urban Geography, 17*(2), 132–151.

Sandercock, L. (1998). The death of modernist planning: Radical praxis for a postmodern age. In M. Douglass and J. Friedmann (Eds.), *Cities for citizens: Planning and the rise of civil society in a global age* (pp. 163–184). Chichester: Wiley.

Schmidt, S. (2004). World wide plaza: The corporatization of urban public space. *IEEE Technology and Society, 23*(4), 17–18.

Sennett, R. (1978). *The fall of public man.* New York: Vintage.

Shields, R. (1989). Social spatialization and the built environment: The West Edmonton Mall. *Environment and Planning D: Society and Space, 7,* 147–164.

Sibley, D. (1995). *Geographies of exclusion.* London: Routledge.

Siegel, F. (1992). Reclaiming our public spaces. *City Journal,* 35–45, (Spring).

Smith, N., & Low, S. (2006). "Introduction: The imperative of public space. In S. Low and N. Smith (Eds.), *The politics of public space* (pp. 1–16). New York: Routledge.

Sorkin, M. (Ed.). (1992). Variations on a theme park: The new American city and the end of public space. New York: Hill and Wang.

Turner, R. (2002). The politics of design and development in the Postmodern downtown. *Journal of Urban Affairs, 24*(5), 533–548.

Valentine, G. (1996). Children should be seen and not heard: The production and transgression of adults' public space. *Urban Geography, 17*(3), 205–220.

Valentine, G. (2004). *Public space and the culture of childhood.* London: Ashgate.

Warren, R. (2002). Situating the city and September 11th: Military urban doctrine, 'pop–up' armies and spatial chess. *International Journal of Urban and Regional Research, 26*(3), 614–619.

Wekerle, G. and Whitzman, C. (1995). Safe cities: Guidelines for planning, design and management. New York: Van Nostrand Reinhold.

Whyte, W. (1988). *City: Rediscovering the center.* New York: Doubleday.

Young, I. (2000). *Inclusion and democracy.* Oxford: Oxford University Press.

Zukin, S. (1995). *The cultures of cities.* Cambridge, MA: Blackwell.

Chapter 5
Expectancy Theory in Quality-of-Life Leisure Indicators – Influences for Integrated Resource Management

Bill Field

Introduction

Resource management planning has undergone significant changes in the level of public participation and decision making during the last few decades. The numbers and varying needs of stakeholders have drastically increased, while methodology and tools to analyze and incorporate sometimes competing demands have only increased piecemeal. It was hypothesized that surveys, based upon expectancy theory, could be used to collect user needs and preference data from under-represented populations to expand integrated planning outreach (Field, 1999). This data would then be incorporated within broader resource management planning processes that focus on satisfying multi-stakeholder quality of life issues.

British Columbia (hereafter BC) has proven to be a good test case as it has undergone significant changes in the way natural resource development activities have been managed in the last two decades of the twentieth century. Resource management planning must now incorporate a level of public consultation and participation not previously seen in BC (Wilson, Roseland, & Day, 1996). Analysis of public participation to date has concentrated on public consultation models involving residents or nearby visitors to the resources in question, stakeholders with economic concerns, and non-industrial public users (Owen, 1998). Lacking in the planning process has been a method of consultation that includes the interests or attitudes of significant user groups, which may not hold local residency. If increasing efforts are to be made to diversify uses, improve quality of experience, and diversify revenues from natural resource development, then monitoring and incorporating the needs of a wide range of user groups in development planning are essential (Robinson, Hawley, & Robson, 1997; Williams et al., 1994).

Expectancy theory has been widely applied to tourism and recreational use planning. Research on recreational users' aesthetic perceptions of the natural environment has revealed the vital importance of previous experience to the formation

B. Field (✉)
International Languages Department, University of Applied Science Landshut,
Am Lurzenhof 1, 84036 Landshut, Germany
e-mail: bill.field@web.de

M. Budruk, R. Phillips (eds.), *Quality-of-Life Community Indicators for Parks,*
Recreation and Tourism Management, Social Indicators Research Series 43,
DOI 10.1007/978-90-481-9861-0_5, © Springer Science+Business Media B.V. 2011

of present attitudes (Manfredo, Driver, & Brown, 1983; Mugica & De Lucio, 1996; Purcell, 1986). Recreational satisfaction research has even more closely linked expectations formation toward recreational outings with previous learning or experience (Schreyer, Lime, & Williams, 1984; Stewart & Carpenter, 1989; Webb & Worchel, 1993). This identification and fulfillment of satisfaction requirements can be directly linked to quality of experiences, and hence to recreational components of quality-of-life (hereafter QOL) indicators. In fact, recent research has directly focused on expectancy theory to evaluate QoL indicators from leisure, and correlations to overall life satisfaction (Neal, Uysal, & Sirgy, 2007).

Recreational managers responsible for providing services to a varied visitor population will benefit in their site management plans by analyzing potential visitor groups' expectations prior to the development of any recreational site. In lands designated for a multitude of human uses, satisfying multiple human demands simultaneously is a significant challenge. Revealing visitors publics' site or quality expectations is an early step in any development plan that aims to incorporate quality of life indicators as one component of success criterion.

Studies have revealed that manager evaluations of site quality or ecological impacts differ from those of users (Shelby & Harris, 1985). Thus, if the needs of various user groups are considered paramount in development design, then the perspective of the user must be adopted into development planning. Evaluative tools need to be developed for gathering and quantifying visitor preference and leisure standards data. Expectancy theory, which has been used by recreational managers for recreational conflict planning decisions, may be adopted for use in gathering such data. For recreational or tourism resources in the development phase, or those that may be undergoing significant alteration, analysis of potential visitors' expectations for site characteristics provides valuable planning data that may be unavailable by other means.

The purpose of this study was to employ an interdisciplinary approach (quantitative research using expectancy theory and qualitative interview-based research) to examine user satisfaction requirements and land and resource management challenges that could be encountered in an attempt to grant heritage status to one of the few remaining larger intact sections of the Cariboo Gold Rush Wagon Road (hereafter the Cariboo Wagon Road). This section of the road, 20 km in length, linked two of the key historical population centers for gold seekers and ends in the representative Barkerville Heritage Townsite provincial park.

Historical Transportation Routes

Background of the Cariboo Wagon Road

The Cariboo Wagon Road was a transportation link between southern BC and the Cariboo goldfields of central BC. Begun in 1861, the road originated in Yale and terminated in Barkerville, at that time the largest community on the mainland of BC (Downs, 1993). Over 400 km in length, this road was to have significant impact on the social and economic history of BC. It transported and supplied large immigrant populations in the central interior of BC and thus forever changed the social and economic organization of the region. Today, little remains of the original roadbed. Neglected after the advent of railways made its transportation services redundant; this road has been paved over or used for other human purposes.

At the time of its completion, the Cariboo Wagon Road was considered the eighth wonder of the world (Downs, 1993). Governor James Douglas had recognized the need for a supply road to the goldfields of the Cariboo, both as a means of supplying the myriad newcomers flooding the region in search of gold and as a method of asserting colonial control over the region (Wade, 1979). In the early 1860s American miners greatly outnumbered colonial residents and represented a threat to the sovereignty of the colonial government. The road was a method of ensuring administrative control and for developing an economy controlled by the crown. This resulted in a system of taxation, judicial governance, and control of the gold resource previously not possible (Patenaude, 1995). Within a few years of the road being completed, Americans were strongly outnumbered by those of British ancestry, and administrative control was firmly in the hands of the Crown.

Heritage Trails and Tourism Management

In the last several decades, the rise of the heritage industry has led to a proliferation of museums of rural and urban life, increasing promotion of historic themed travel, and incorporation into the tourist industry of historic landscapes and relict features of agriculture, industry, and commerce (Butlin, 1993; Weiler & Hall, 1992). The advent of increasing interest in, and development of, heritage sites requires significant investments in management planning that recognizes the complex values and priorities of land use decision making.

The growth in heritage tourism is attributed to a growing awareness of heritage, greater affluence, increased emphasis on leisure, greater mobility of travelers, and increasing overall tourism visitation rates (Zeppel & Hall, 1992). The variety and type of heritage attractions available attract a diverse range of visitors, with many attractions being popular with both resident and non-resident visitors. Depending on the reasons for visitation and the product offering of the heritage resource, sites can have a diverse range of visitor types and reasons for visitation or a narrow, specialized type of visitor. An increasing inventory of heritage attractions results in the discerning heritage visitor having a range of product offerings and at times multiple offerings of a similar historical emphasis or period representation.

As the global demand for heritage resources increases, many communities have begun to use historic preservation as a method of attracting tourists in order to diversify rural economic development (Go, Milne, & Whittles, 1992; Province of BC, 1992). Indeed, both the province of BC and the Federal Government of Canada, amongst others, have emphasized the historical aspect of Canada's various cultures, structures, and natural landscapes in tourism promotion (Province of BC, 1993a). By offering a diverse range of historical representation, heritage managers are not only able to preserve features of noted heritage significance, but provide diverse tourism resources to a discerning public. This preservation and economic diversification positively impacts multiple QoL indicators, such as social, leisure, and employment needs.

BC government tourism marketing staff has identified heritage tourism as of significant interest to varied visitor groups in BC (Province of BC, 1991). Within BC, communities and provincial and federal governments are now reviewing the heritage resources in their area and are identifying cultural resources that may be exploited as a source of economic development. This is aptly illustrated in the promotion of First Nations tourism as a central underpinning of Super Natural BC tourism campaigns, but also has significant importance to other facets of BC tourism. Both Barkerville and Wells are prime examples of the promotion of an industrial heritage (i.e., gold mining) as the primary element to local tourism development. The late 1800s history of silver and gold mining in southeast BC has also begun to be promoted as significant to the historically distinguishing features of the areas culture and development (Bowers, 1998).

Within the heritage tourism industry, heritage trails have proved a popular form of tourism development throughout the world (Boniface & Fowler, 1993; Yale, 1991). Visits to heritage trails offer a combination of exploring nature and gaining an educational or heritage experience. Internationally, heritage trails have been seen as an effective method of linking museum sites, local communities, and heritage resources that on their own might not draw heavy visitation. Heritage trails aptly bridge the gap between cultural and natural representation and have stimulated rural economies that were unable to attract large visiting publics with a previous narrow range of tourism products.

The setting of heritage interpretation in a modified, but largely natural environment has proven attractive to many tourists (Wright, 1996). In Canada, there are a vast variety of trails or causeways which can be considered to be of historical significance, including First Nations trading routes, industrial causeways, on both land or water, and early colonial exploration routes (Ministry of Canadian Heritage, 1994). Within BC, heritage trails include, but are not limited to, the following: trade routes used by First Nations peoples, the Grease Trail (the First Nations designation of the Alexander MacKenzie Trail) being a prime example; colonial trails which opened regions to economic exploration such as the Cariboo Wagon Road, which this work examines; and the West Coast Trail, a life-saving trail for shipwrecked sailors on the west coast of Vancouver Island.

Heritage trails pose unique challenges to Canadian heritage managers, as they often transit vast areas while occupying only a narrow strip of land recognized as a heritage resource. Often a myriad of connected heritage resource sites may also be found along such trails within a diverse resource and land ownership structure. Previously, provincial or federal heritage legislation did not make provision for protection of landscapes as heritage resources (Roszell, 1996). Recognition of earlier limitations in heritage policy has resulted in more expansive definitions and responsibilities for heritage protection with an emphasis on increased representation in the economic or social history of Canada (Ministry of Canadian Heritage, 1994).

Purpose and Design of the Study

The purpose of this study was to develop a method of surveying heritage visitors' expectations for certain trail characteristics that could be encountered, as well as to document potential satisfaction levels for multi-user contacts, if they were to travel on a Cariboo Wagon Road heritage trail. These characteristics as addressed in the survey included recreational encounter types and noise and viewscape impacts from industrial uses including forestry and mining. A secondary purpose was to evaluate the manner in which various stakeholder parties involved in sub-regional planning that include local trail systems could incorporate visitor data into planning decisions that impact resident and non-resident quality of life indicators. This was achieved through a series of interviews with primary stakeholders in land planning processes in the region through which the proposed heritage trail transits.

Diverse economic and social uses of these historically significant landscapes, combined with the rights of private land titles, can result in a combative and ultimately unsatisfactory land and resource planning process. The protection of linear heritage trails on such lands requires new, innovative approaches to heritage preservation. Land and resource planning systems which can accommodate multiple interests and parties while encouraging sensitive multiple uses of the landbase are attractive to governments sensitive to community quality of life issues within resource planning.

More open, public processes of land planning do not automatically ensure representation for all users of a natural heritage resource. The nature of tourism or recreation means that visitors may come from long distances to enjoy the benefits of the natural resource. This distance can pose significant barriers to the inclusion of their viewpoints in planning decisions. If visitors are without a voice, then their concerns cannot be addressed when land use decisions need to be made. When visitors are unhappy with the results of these decisions, a likely consequence is dissatisfaction with the experience, a consequent communication of these feelings to other potential visitors, and future displacement of tourism visits. Such outcomes will adversely impact QoL levels for both communities dependent on tourism revenues, as well as visitor satisfactions. The collection of data that identify the visiting publics desires for the appearance and management of natural heritage landscapes aids both the planner and the manager of the landbase in question.

The Quantitative Methodology

A survey research method was selected for this study due to the large size of the visitor population under study, the ability to analyze the variables of concern through a sample population, and the additional cost of other research methods. A self-administered mailout survey was selected for the following reasons: ability to collect a representative sample in the research time allotted, the concern over non-response on site due to potential constraints upon the respondents (time or weather), and the elimination of interviewer bias.

The advantages of a mail survey are its ability to reach the intended audience and the relatively low cost involved. Disadvantages include relatively low response rates and lack of additional feedback if the survey design proves difficult for the audience to comprehend. Thus, low response rates may be lack of interest or lack of comprehension, yet the researcher will be unaware of reasons for low response. Pre-testing of the survey design may help to overcome this concern (de Vaus, 1991).

Addresses for the mailout survey were collected by the author and a research assistant. Procedures for address collection and all materials used were standardized, and training in these procedures for the research assistant occurred on site, and were administered by the author. All data collection materials used onsite, which included the introduction speech, the address collection sheet, an explanatory map outlining the trail location for potential respondents and concluding remarks, were developed by the author.

The participant intercept took not more than 3–4 min from initiation of the contact to completion. The information required was confirmation of willingness to participate, sex of respondent (filled in by interviewer), number of previous visits to the site in the last 3 years, and mailing address; residency information was taken from the supplied address. The final desired sample size was approximately 300 individuals completing questionnaires. This would provide for sub-sample populations of a minimum 50 based on up to four residency codes (local, other BC, rest of Canada, and international). Adams and Schvaneveldt (1991) and de Vaus (1991) contend that minimum sub-samples from an N of 30–50 are necessary in terms of just statistics. Similar relevant survey research would indicate that non-response rates up to 50% are common with mailout surveys (Baker, Hozier, & Rogers, 1994; Dillman, 1978; Fowler, 1995). Thus, a total sample size of 600 adults was likely to provide adequate sub-sample sizes. The study survey had a final response rate of 76.6%, a substantial rate that was perhaps influenced by the on-site participant selection and a series of two reminder letters.

Respondents

The site of the respondent address collection was Barkerville, a provincial historic townsite park that receives approximately 100,000 visitors over a 12-week period during the summer (Jim Worton, Manager, July 31, 1997, Barkerville Historic Townsite, personal communication). The respondents were systematically selected by approaching every eighth adult site visitor and requesting that they participate in a mailout survey. To be eligible, respondents had to be of at least age 16, able to communicate in the English language, and express a willingness to participate. If the eighth person selected was ineligible due to age, then the interviewer selected the following adult in the same party. If this person declined to participate then the interviewer returned to asking every eighth adult. The approach site for participation in the survey was at the entrance/interpretation center, through which people had to transit in order to enter or exit the site. The center provided the advantage of being under cover during inclement weather. Non-participation rates for the survey were

tracked by seven non-response codes and by weather conditions on the address collection date. The onsite refusal rate to participate in the survey was 35.6%. Overall
the most often cited reason for non-participation was "not wanting to participate," or
with "not being competent in English" being the next most frequently cited reason.

Instrumentation

The questionnaire included 29 questions with 89 items (see Appendix 1 for
some sample questions). The questionnaire was predominately composed of close-
ended questions on encounter types, demographic information, and expectations of
impacts from industrial activity. Open-ended questions pertaining to personal preferences for recreational experiences and certain viewpoints toward industrial use
(noise and viewscape effects) were also included to ensure that respondents were
able to adequately respond to attribute characteristics using their own descriptive
responses.

The questionnaire was designed with three dependent variables: recreational user
encounters, impacts from noise, and impacts from viewing resource extractive activities (logging and mining). Respondents were asked their expectations if they were
to use a heritage trail historically and geographically linked to Barkerville. The
independent variables included residency, importance of heritage tourism to the
respondent (a measure for perceived satisfaction levels), sex, and education.

Previous recreation surveys that explored respondents' expectations for a recreational experience have targeted issues of crowding, need of solitude, landscape
preferences, performance expectations, or satisfaction identification in the tourism
industry (Baker & Fesenmaier, 1997; Johnson, Anderson, & Fornell, 1995; Schreyer
et al., 1984). While useful for supporting literature and design of certain questions, these other surveys proved inadequate for exploring the specific relationship
examined in this research.

The survey comprised three specific sections. It was designed so as to identify
overall heritage knowledge and importance, to assess expectations for trail design
and interactions, and to provide demographic and interest group data. Section A of
the questionnaire included three questions that inquired about respondents' previous visits to Barkerville, other heritage sites, and previous visits to heritage trails
and museum or cultural center with an attached heritage trail. The final question
provided valuable information to Barkerville Heritage Townsite management when
exploring the connection of heritage trails to museums. A simple definition of heritage trails was provided for the respondent: "heritage trails were defined as routes
that show us how our ancestors or previous cultures lived and worked (e.g., trading
networks, exploration routes, or transportation links such as railways, waterways, or
roads, amongst others)."

Section B was designed to reveal respondents' attitudes and expectations regarding a variety of trail conditions. This section proposed a series of conditions that
might be found on the trail and asked for respondents' attitudes to each of these
conditions. The questions gauged attitudes toward recreational encounter types and

noise or viewscape impacts from logging or mining. The importance of visiting historical sites to the respondents overall interest in tourism was one of the expectancy level variables under investigation.

Section C collected information on the participants' experiences at Barkerville, and general demographic information to be used in categorizing sub-groups. These data were used to reveal expectation levels for the respondent's site visit experiences and to segment visiting populations. A variety of questions were included to determine expectancy levels of respondents.

The survey was pre-tested on a group of university students as a convenience sample to maximize validity and reliability. The students were instructed to comment on their comprehension of the survey, order of questions, clarity of language in the questions, whether multiple responses were possible for any question, and the clarity of the instructions for the survey. Results indicated that the wording of several questions was unclear as well as clarity of choices between responses for several questions. Survey modifications were made as deemed necessary.

The Qualitative Methodology

The purpose of the qualitative research was to identify how visitor expectations for trail design could be incorporated into QoL standards for various stakeholder parties, and subsequently used in an integrated planning process. This research comprised a series of interviews with identified planning participants.

Interview Procedures

Seven semi-structured interviews were conducted with individuals representing the primary stakeholders in the planning process revolving around lands bordering the proposed heritage trail. These interviewees consisted of a heritage manager, board member of the local historical society, board member of the local recreation and tourism association, forest industry representative, mining representative, and regional land planning officials (represented by various departments within the BC Forests Ministry).

All individuals were identified as significant to the ongoing planning process and had a history of participation in previous regional planning processes. The heritage manager was directly responsible for managing heritage resources connected to the Cariboo Wagon Road. The historical society is the primary non-government organization regionally involved in heritage preservation efforts. The local recreation and tourism association develops trails in the region and promotes tourism development. The forest industry representative is responsible for operations on lands that the Cariboo Wagon Road traverses. The forest ministry officials are responsible for harvest plans that occur in the region with some members also acting as the responsible agency for regional land planning. The mining representative is an employee of the energy and mines ministry responsible for planning participation and familiar with regional placer mining operators. Survey respondents were also specifically asked their opinion of the importance of these planning participants' opinions when developing a trail management plan.

Interviewees were supplied with a copy of the survey that was utilized for quantitative data, a report on the results of the survey prepared by the author, a covering letter explaining the nature of the research, and a list of 11 questions (see Appendix 2). Participants were asked to respond to the questions and were also provided with sufficient time to expand upon any topics they felt pertinent to the issues at stake. All interviews were in person, or in limited cases of availability, through telephone communications.

Participants were notified that they could refuse to continue participation at any time and could refuse to answer any questions they chose. All conversations were taped and interviewees were notified of this fact. Again, interviewees were informed that they had the right to request that their session not be taped, and this would be respected. No interviewee requested this. Interviewees were informed that all conversations were anonymous, and no individual would be identified in any way in the thesis documentation. All documents pertaining to this study passed an ethics review process prior to the onset of the fieldwork.

The interviews were analyzed with the use of content analysis. Content analysis is considered a set of methods for analyzing the symbolic content of any communication (Singleton, Straits, & Bruce, 1993). Responses were coded into principal themes, and then further analyzed into sub-themes. The use of a semi-structured interview format allows the author to develop content categories by relating the characteristics of responses against questions being asked. Thus, patterns of responses and critical words and subjects were assigned separate categories and analyzed by the role of the interviewee as stakeholder.

Delimitations and Limitations of Study

Several delimitations were identified in this study. In Canada, limited research has analyzed heritage trail visitors' attitudes to trail management practices. A delimitation of this study is the representativeness of off-site survey responses to hypothesized QoL impacts (i.e., in this stuffy satisfaction levels) from proposed recreation or industrial impacts. Onsite personal experiences may differ from an individual's expected impacts. The study was also restricted to heritage visitors; other visitors to the region may have different QoL satisfaction from the benefits of a heritage trail. In addition, not all planning stakeholders may have participated in the study interviews, thus qualitative responses are restricted to those participating in this study. Lastly, this survey was produced only in the English language, thus responses can only be considered to reflect the expectations of visitors competent in the use of the English language.

Respondents surveyed at the heritage site were composed of the so-called rubber tire tourism (either bus or rental/personal vehicle) and may not represent visitors who were capable of traveling or recreating on a 20 km heritage trail. The level of physical fitness of visitors to Barkerville may result in lower numbers of potential trail recreationists than would a study undertaken on an existing heritage trail. The present trail infrastructure allows for a variety of transportation modes to be utilized including horse, bicycle, motorized transport, or self-propelled options, which suggests that most visitors could find an acceptable method of recreating on the trail.

Hypothetical questions about future behavior cannot necessarily reproduce behaviorally relevant issues (Fowler, 1995). In order to compensate for the challenge of inquiring as to respondents' future expectations the survey design incorporated both closed and open-ended questions. To the extent that questions about a hypothetical situation can build on respondents' relevant past experiences and direct knowledge, the responses will be more accurate (Fowler, 1995; Mugica & De Lucio, 1996). The direct historical connection of the onsite survey location with the trail under study should generate accurate responses as these visitors have already expressed interest in heritage attractions by their attendance at Barkerville.

Expectancy Theory and Visitor Satisfaction

Recreational or tourism planners provide experiential services to users or visitors of recreational and tourism facilities or sites. It is often argued that the attitudes, expectations, and perceptions of tourists or recreationists are significant variables in the individual setting of goals, visitor behavior, and the final satisfaction of the individual (Ryan, 1995). In order to optimize satisfaction, managers and planners attempt to discern and then match visitor desires or preferences with site characteristics. One common method of evaluating these preferences is through identifying previous similar product or event experiences that potential visitors may have had and matching those interests that are found to be compatible with the available resources offered (Schreyer et al., 1984). The potential for success or visitor satisfaction is highest when the expectations of potential visitors are aligned with the actual site or resource characteristics being offered. The present experience is thus not isolated from previous experience, but forms a foundation for how the visitor will evaluate an ongoing experience.

Research into visitor or consumer satisfaction is not new. Interpretations of the causes and effects of satisfaction have had significant implications for managers and QoL researchers in various fields. Marketing research on consumer satisfaction has yielded valuable models on the antecedents and consequence of expectation formation and its interpretation by both the service provider and the consumer (Oliver, 1980; Ryan, 1995).

Understanding the consumer decision process and how expectations are developed and defined by the individual is a goal of managers in aiding resource allocation decisions (Manfredo et al., 1983). The direction of satisfaction research has focused on individual expectation formation and the ability to discern group behavior from individual expectation. Such research has expanded the use of expectancy theory into many fields. The leisure or tourism fields are extremely well suited to this approach, as they are primarily experiential in nature, and the consumed experience becomes the service by which satisfaction is judged. As leisure satisfaction is a critical QoL indicator, research to quantify and differentiate user types can be critical in establishing diverse need and performance levels.

Human Behavior and Expectancy Theory

Humans often tailor their behavior or actions to reflect their expectations of expected future outcomes relative to their efforts. Olson, Roese, and Zanna (1996) state that every deliberate action that humans undertake rests on assumptions about how the world will react in response to their action. In social psychology, this concept of human expectations has become known as expectancy theory. Expectancy theory states that expectancies are beliefs about a future probable state of affairs, which links the "future with an outcome at some level of probability ranging from merely possible to almost certain" (Olson et al., 1996, p. 211). This probability is affected by a variety of factors, some within the individual's control and others external individual influence, but both sets of factors influencing motivation to achieve a positive QoL need.

Recreation Behavior and Expectancy Theory

Numerous studies on recreation behavior moderated by past experience have high-lighted this connection of motivation to previous learning (Manfredo et al., 1983; Mugica & de Lucio, 1996; Schreyer et al., 1984). There is increasing inter-est in understanding the link between information available and leisure behavior. Motivation research has examined the critical linkages between the cognitive state of the individual (attitudes, feelings, and motives) and recreational behavior (Ewert, 1993; Schreyer et al., 1984).

Recreation surveys based on expectancy theory have primarily been utilized for research into issues of crowding, service satisfaction, and desire for solitude (Stewart & Carpenter, 1989; Webb & Worchel, 1993). Tarrant, Cordell, and Kibler (1997) found that environmental conditions, such as the number and type of other users, location, and activity, influence crowding levels by changing recreation-ists' expectations about the social situation. Pre-visit knowledge of site conditions would likely reflect on changes in behavior to avoid less desired encounter types or frequencies. Expectations must relate to actual conditions in order for avoid-ance behaviors to be successful. This supposition has been supported by studies that illustrate that recreationists seek relevant information prior to their trip activity (Manfredo et al., 1983; Stewart & Carpenter, 1989). The need for expected condi-tions to match actual conditions is necessary in order that the individual is able, in future, to act upon information to achieve a desired QoL outcome (e.g., a positive recreational event).

Expectancy theory may be a useful tool for designing a survey that will predict attributes a visitor group seeks or highly values when planning to visit a heritage trail. The use of a survey based on expectancy theory to reveal data that can be incorporated in facilities planning is an innovative application of this theory. Ewert and Hood (1995) contend that different wilderness areas attract visitors desiring or expecting different types of wilderness experiences. This contention raises the question of managing for homogenous values or employing a homogenous management regime, when a large degree of variability may be found among users (Ewert, 1998). Some of this variability will be the effect of an individual's investment in pursuing the particular activity. This variability reveals the necessity of surveying visitors' expectations as to what the product attributes should be for a heritage trail they expect to utilize.

Recent recreation research has explored the relationship of respondents' expectations of a recreation experience to that found onsite. Once the motivation to recreate or undertake a tourism adventure has been applied, expectation, in turn, will determine performance perceptions of products and services as well as the perceptions of satisfaction (Gnoth, 1997). Previous experience is linked to a recreationist's expectations; this use of previous knowledge to affect future perceptions or decisions is the fundamental component of independent action in expectancy theory (Olson et al., 1996). The role of previous experience in landscape preference or environmental assessment has recently been extensively researched (Manfredo et al., 1983; Mugica & De lucio, 1996; Shelby & Harris, 1985).

Proximity relationships to sites visited have also been explored and used in this study as an independent variable, to evaluate expectancy development. Ewert and Hood (1995) have employed the use of expectancy theory to evaluate recreationists' expectations and perceptions of the wilderness experience by their residential proximity or lack of, to the recreational site visited. The rationale is that since visiting urban-proximate (UP) recreational locales requires less travel effort, time expenditure, planning effort, and financial cost, than does travel to remote locales, this reduced effort will be evident in lowered expectations about site quality (Ewert & Hood, 1995; Ewert, 1998). Different visitor segments should evaluate outcomes in differing fashions.

This link to positive outcomes and the investment in the activity has interested tourism and recreation managers for some time. Much research has occurred on segmenting the travel market by benefits desired or realized (Baker et al., 1994;). The goal in this strategy is to match the product desired and the eventual outcome and to attract the most desired visitor population. The Province of BC (1991; 1993a) targets overseas visitors as an attractive tourism market due to their lengthier stays and the higher levels of expenditures they incur, in comparison to Canadian visitors. An analysis of benefits desired for European visitors highlights the attraction of a natural environment, cultural history (including First Nations), and the urban environment (Province of BC, 1991). Thus, provincial tourism marketing programs are fabricated on viewing non-resident (and more specifically European) visitors or recreationists as desirable due to their increased spending over local residents. The increased investment in reaching these destinations should directly effect visitors desired satisfaction and, hence their QoL values for the experience.

This travel investment and residency variable formed a basis for the key independent variable; it was hypothesized that non-resident visitor would have a more significant investment in reaching Barkerville or the proposed heritage trail, and thus should hypothetically report less tolerance for negative impacts. It was expected that the distant visitor would report a lengthier period of planning prior to site arrival, and higher importance of heritage as a leisure value. This hypothesis is generally supported by market research that indicates that regional visitors are far more likely to be "visiting friends or family" than other visitors, and these visitors report substantially lower rates of visiting national, provincial, or historic sites than those market segments not visiting "friends or family" (Research Resolutions, 1996). In addition, US and overseas visitors indicate greater interest in natural, cultural, and heritage tourism than do BC or Canadian residents (Research Resolutions, 1996).

Expectancy Theory and Travel

Figure 5.1 illustrates the antecedent conditions affecting expectation formation. It outlines the three factors that interact to develop the motivation to act: previous experience, level of control and investment in the activity by the individual, and an assessment of the overall environment.

The impetus to undertake the behavior or action is the expectation that action results in desired outcomes. This belief is the primary motivator to act. When the subsequent (the result) matches the antecedent (the expectation) then satisfaction

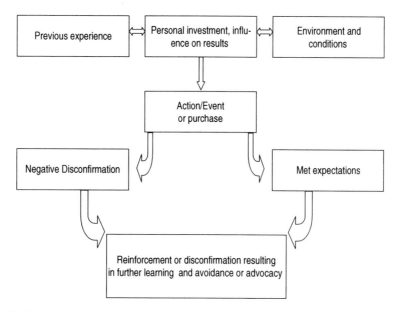

Fig. 5.1 Expectancy theory and outcomes

occurs and the individual feels justified in their behavior. The individual's behavior is reinforced by this positive outcome. When the result does not match the expectation then disconfirmation occurs. That is, the individual encounters the dissonance of unmet expectations and seeks an explanation for this result. Both unmet and met expectations result in the further development of previous experience; this experience then becomes the antecedent to future actions.

For managers, the ability to first identify expectations held by potential visiting groups and then supplying the information that would justify the commitment to visit is a critical component of matching service to expectations. The motive to act will then be a function of previous experience with the situation, an affective component of the individual's history with a similar situation, and a classification of the present situation in comparison to the schema. Matching these components of the cause for action to the product offering is the responsibility of management. A successful outcome results when dissonance or disconfirmation does not occur and the individual considers his/her expectations to be met or exceeded by the activity or resource.

The contention of this work is that heritage visitors, in part, will evaluate a particular setting by the types of expectations they hold regarding that setting. That is, how visitors judge the quality of a site or an activity is a direct result of their expectations about features they value and expect in leisure. The planning that they put into undertaking the activity then becomes a judgment about their confidence that the event will meet their expectations. Recently, significant research has focused on how visitors form their expectations of product settings and quality and how these expectations affect subsequent satisfaction levels (Cho, 1998; Duke & Persia, 1994; Ewert, 1998; Ryan, 1995). This recent research has differentiated between two aspects of previous experience crucial in the expectancy formation: the actual activity or similar site experience that individuals use to develop their expectations about similar products/conditions, and the messages and communications that individuals have received about their intended activity or site being visited. The latter aspect is developed through intentional or unintentional marketing efforts of the tourism and recreation industry. Both factors will play an active role in expectancy formation and subsequent satisfaction/dissatisfaction outcomes.

Integrated Resource Management Planning

One of the primary collaborative resource management approaches adopted in BC has been the Land and Resource Management (LRMP) planning system. This system encourages planning that emphasizes community sustainability through participation and management control that benefits multiple QoL indicators such as environmental, socioeconomic, and ecological dimensions (Owen, 1998). The broad public acceptance of land and resource management policy objectives is an inherent goal of LRMP. A primary aim of planning is to balance demand and capacity so that conflicts are minimized and resources are most effectively utilized with the least deterioration of the resource base (Pigram & Jenkins, 1994). Encouraging public participation both provides a voice to the public as to their specific needs and provides a rationale for the future. Owen (1998) identified the publics' concern that government or industry lacked a comprehensive plan as to how the forestry resource would be maintained over the long term as instrumental to the overwhelming demand for broader planning participation. As these public parties have gained legitimate rights of access to publicly owned lands they have stimulated the push for inclusion of public user groups in the initial stages of planning rather than after resource allocation decisions have been made. The recognition that all parties with a legitimate stake in an issue must be invited to participate in planning decisions is a foundation of LRMP and one of the critical dimensions of IRM planning. Duffy et al. (1998), in an analysis of shared decision making for land and resource planning, differentiated between two forms of stakeholder negotiation: positional negotiation where participants define themselves by needs, desires, or concerns, and interest-based negotiation where participants recognize the difference in values and seek a common ground upon which to seek out agreement. These two forms of negotiation define the commonly accepted bi-polar negotiation types. The former is a game of win or lose, while the latter is focused on win–win or an acceptable middle ground (Institute for Participatory Management and Planning, 1994). The latter option does not preclude one group appearing to gain a more advantageous decision in comparison to another, but also suggests that accommodation can be made. The result is often a modification of one desire for another, such as preservation for one area while allowing harvesting in a different region, often focused upon maximizing conditions such as QoL values. LRMP approaches are very amenable to this type of negotiation. This has already been successfully followed in the Wells-Barkerville area where forest harvest practices have been modified near popular cross-country ski trails in order to limit viewscape impacts, while lesser desired recreational areas have seen increased harvesting.

Incorporation of public or user attitudes within resource management plans is also desirable as a planning tool to more accurately assess where limited management resources can best be applied. Significant attitudinal research to date on recreational visitor attitudes has indicated that often managers may hold different expectations toward visitor attitudes or needs than actually exist (Bright, 1997; Vining, 1992). Attitudinal data collection for potential heritage trail visitors can provide valuable QoL leisure variable information that may assist in scoping resource management decisions. Previous research has indicated that when the public is consulted, has an active voice in, and receives regular communication of ongoing planning processes they are more supportive of the eventual outcomes and report higher satisfaction levels (Penrose, Day, & Roseland, 1998; Robinson et al., 1997). As public planning participation is still in its infancy, methods of inclusive planning focused on various QoL indicators must still be developed and evaluated (Institute for Participatory Management and Planning, 1994).

This combined format of resource development activities with leisure or tourism is common in Europe (Aronsson, 1994; Phillips & Tubridy, 1994). Until recently in Canada, public lands were often primarily dedicated to single use purposes. Secondary users of crown lands had little or no voice in planning matters (Owen,

1998). With increasing demands on public lands, the responsible governing bodies have now turned to more inclusive planning processes in an effort to extract maximum QoL values for use of public lands, with the least detrimental impacts.

Quantifying the expectations of the visiting public will aid in the acquisition of presently scarce technical and social data on development of the trail and by recognizing the expectations of various planning stakeholders will reduce some of the inherent LRMP challenges. Throughout this study, it will become evident that expectations of all parties form the foundation of either planning conflict or positive QoL realization.

Data Analysis

The analysis portion of this study is separated into two sections. The first section is a quantitative evaluation of heritage site visitors' attitudes toward recreational and industrial uses of the land surrounding a proposed heritage trail. These data were evaluated using both non-parametric and parametric statistics to assess expectations about potential impacts. The second section is a qualitative examination of how planning stakeholders view the information collected in the survey, how said information may reflect acceptable QoL outcomes, and how planning stakeholders envision visitors' attitudes being incorporated within the final land planning implementation stage.

Screening the Data

Coding of the survey data was carried out after initial review of the first mail-back replies. Open-ended questions had numerous response codes, including an "other" category for non-identified responses. Non-dichotomous close-ended questions were coded on a Likert scale of 1–5, other than limited questions that did not employ a Likert scale.

The taped interviews with regional land planning stakeholders were transcribed by a research assistant using standard transcribing procedures. Coding of the transcriptions was based on content analysis. Responses were coded by theme and sub-theme variables identified in the transcription. Interviewees had the opportunity throughout the interview process of adding additional information, or diverging from the interview format at any time. Each participant also received a copy of the interview transcription for their verification as to the text content and allowances were made for further input. This did not result in any additions or challenges to the transcript records.

Quantitative Data Analysis Highlights

Data analysis revealed surprising exceptions from commonly held beliefs of expectancy research. Previous research, and this study, postulated that expectations increase with trip investment, demographic characteristics of visitors could differ by residential proximity, and that tolerance for industrial activities would differ by investment level (trip effort). The majority of these hypotheses were not validated by the results.

Table 5.1 provides a breakdown of respondent demographic and trip characteristics. Age and education variables were tested by residency with one-way ANOVA and post hoc analysis. USA residents proved to be significantly older than international visitors ($F(4, 433) = 4.071, p = 0.003$) and USA residents had significantly more education than local or other BC residents ($F(4, 429) = 6.607, p = 0.000$). Age and education met the assumption of homogeneity of variance (Levene statistic).

The travel motivations of Barkerville visitors were evaluated by assessing the "importance of visiting heritage sites in visitors travel plans" by residency. As can be seen in Table 5.2, international visitors placed higher importance on heritage than local visitors ($F(4, 426) = 2.659, p = 0.032$). Importance of heritage met the assumption of homogeneity of variance. This finding supported a study hypothesis that distance would affect expectancy and satisfaction need levels.

Table 5.1 Characteristics of the respondents

	Local ($n = 131$)	Other BC ($n = 214$)	Other Can ($n = 29$)	USA ($n = 23$)	Intl. ($n = 41$)
Age					
Mean	40.7	43.2	43.7	49.0[b]	37.7[b]
Education (years)	($n = 131$)	($n = 214$)	($n = 29$)	($n = 23$)	($n = 41$)
Mean	13.2[b]	14.0[b]	14.2	15.9[b]	14.9
Sex					
Female %	54.5	50.9	58.6	39.1	40.0
Male %	45.5	49.1	41.4	60.9	60.0
Trip characteristics					
Plan to visit?	($n = 129$)	($n = 214$)	($n = 29$)	($n = 22$)	($n = 41$)
Yes (%)	84.5	85.5	62.1	50.0	65.9
No (%)	15.5	14.5	37.9	50.0	34.1
Length of trip	($n = 120$)	($n = 210$)	($n = 29$)	($n = 23$)	($n = 40$)
Days (mean)	2.9	11.1	16.6	31.4	27.3
Days (median)	1.0	8.0	12.0	15.0	27.0

[a]One-way ANOVA significant at $p < 0.05$.
[b]One-way ANOVA significant at $p < 0.01$.

Table 5.2 Importance of heritage to overall travel plans

	Local	Other BC	Other Can	USA	Intl.
Importance of heritage					
Mean	3.12[a]	2.84	3.13	2.77	2.53[a]
SD	1.12	1.13	1.24	1.34	1.07

[a]One-way ANOVA significant at $p < 0.05$.
[b]One-way ANOVA significant at $p < 0.01$.

Survey respondents were asked their attitudes toward a variety of potential recreational activities that could occur on the heritage trail, as well as their likely season of use and personal preference for recreational choice (see Table 5.3). They were queried as to whether they support or oppose a range of recreational uses on the proposed heritage trail.

As can be seen from Table 5.3 similar responses were found for all groups with the exception of cross-country skiing, where international visitors reported less support for the activity than all other populations ($F(4,427) = 10.957$, $p = 0.000$). The variable cross-country skiing met the assumption of homogeneity of variance. Overall, the hypothesis that support for different recreational encounters would differ between populations (distant proximity would have lower tolerance for other recreational types) was not supported by the data.

The preferred type of recreational trail use was further supported by a question that asked the preferred season of trail use. Results indicated that summer was the most frequent single season choice (24.9%) while those indicating multiple seasons that included summer represented 77.3% of responses. The choice of winter or multiple seasons of winter and other (excluding winter and summer at 3.8%) was only 2.3%. These results suggest limited involvement in winter activities, and thus, lower conflict encounters from winter use of the trail.

Respondents were asked a series of questions about potential impacts of industrial activities near the trail. Visitors universally opposed mechanized logging, mining on the edge of rivers, and hunting (see Table 5.4). However, horse logging received a neutral response from all residency groups. Tourism received the strongest support from survey respondents. Results that were significant included international visitors having significantly less opposition to mining on rivers than

Table 5.3 Support for recreation activity

	Local ($n = 130$)	Other BC ($n = 211$)	Other Can ($n = 29$)	USA ($n = 22$)	Intl. ($n = 41$)
Hikers	1.32	1.29	1.32	1.36	1.51
	SD 0.625	SD 0.551	SD 0.475	SD 0.581	SD 0.637
Horseback	2.22	2.26	2.37	2.00	2.29
	SD 1.00	SD 1.02	SD 1.04	SD 1.06	SD 0.873
Horse and wagon	2.44	2.35	2.48	2.09	2.41
	SD 1.05	SD.987	SD 1.02	SD 1.09	SD 0.893
Bicycle	2.48	2.63	2.62	3.77	2.92
	SD 1.10	SD 1.12	SD 1.04	SD 1.17	SD 0.877
Motorized summer	4.31	4.47	4.41	4.27	4.19
	SD 0.997	SD 0.679	SD 0.682	SD 1.03	SD 0.980
Skiers	1.78[b]	1.87[b]	1.82[b]	1.63[b]	2.78[b]
	SD 0.819	SD 0.929	SD 0.710	SD 0.789	SD 1.10
Snowmobile	4.09	4.08	4.41	3.95	4.24
	SD 1.13	SD 1.06	SD 0.824	SD 1.13	SD 0.916

1 = strongly support; 5 = strongly oppose
[a]One-way ANOVA significant at $p < .05$
[b]One-way ANOVA significant at $p < 0.01$

Table 5.4 Human activities on or near the trail

	Local (n = 130)	Other BC (n = 211)	Other Can (n = 29)	USA (n = 22)	Intl. (n = 41)
Mechanized logging	3.68	3.80	4.06	3.73	3.70
	SD 1.22	SD 1.13	SD 1.03	SD 1.25	SD 0.966
Horse logging	2.22	2.40	2.75	2.52	2.52
	SD 0.932	SD 1.01	SD 1.12	SD 1.08	SD 0.876
Mining on rivers	3.56	3.76[b]	3.65	3.56	3.02[b]
	SD 1.18	SD 1.09	SD 1.17	SD 1.37	SD 0.986
Guided tourism	1.90[a, b]	2.07	1.89	2.54[b]	2.32[a]
	SD 0.792	SD 0.881	SD 0.817	SD.911	SD 0.888
Hunting activity	3.54	3.79	3.92	3.39	4.07
	SD 1.23	SD 1.17	SD 1.01	SD 1.26	SD 1.07
Non-guided recreation	2.07	2.25[b]	2.17	1.69[b]	2.55[b]
	SD 0.985	SD 1.02	SD 0.848	SD 0.634	SD 1.33

1 = strongly support; 5 = strongly oppose
[a]One-way ANOVA significant at $p < 0.05$
[b]One-way ANOVA significant at $p < 0.01$

other BC visitors ($F(4,423) = 3.574$, $p = 0.007$), USA visitors having significantly more opposition to guided tourism than local visitors ($F(4,429) = 4.154$, $p = 0.003$), and USA residents having significantly more support for non-guided recreation than international residents ($F(4,428) = 3.184$, $p = 0.014$). "Mining on rivers" and "guided tourism" met the assumption of homogeneity of variance. Non-guided recreation did not meet the assumption of homogeneity of variance; the Games–Howell test indicates that the significant difference between populations for "non-guided recreation" lies between the USA and other BC ($p = 0.006$) and between the USA and international ($p = 0.010$). Overall, using "importance of visiting heritage sites to overall interest in tourism" as an independent variable did not reveal differences in responses between groups.

Open-ended responses to this question (i.e., why respondents felt certain activities should be allowed on or near the trail) indicated that low-impact or historical activities were the most desirable. Responses also suggested that horse logging is considered both low impact and of a traditional nature. Respondents did not statistically differ in their responses by the independent variable "importance of visiting heritage sites to overall tourism interest."

Questions pertaining to the impacts of active industrial activities were viewed as being negative toward overall respondent enjoyment of the trail. Surprisingly, international visitors cited statistically less negative impact than other respondents from a variety of industrial activities, including "meeting logging/mining workers," "hearing flying aircraft," "seeing logging activity near streams/lakes," "seeing clearcuts within 100 m of the trail," "mining that would occur in gravel banks along streams" than most other respondents. Many of the results met the assumption of homogeneity. The higher acceptance for some industrial or historical–industrial activities by international visitors may be a result of less exposure and, thus, knowledge of such activities.

Qualitative Data Analysis Highlights

All stakeholders stated that non-residents rarely have a method of participating in resource planning due to barriers of distance and time commitments needed in order to participate. Due to this constraint, while the right to participate exists, the likelihood of non-residents participating is considered small, and thus collecting information through use of a survey and then having a stakeholder with a vested interest in non-residents' attitudes represent these concerns was considered adequate by all parties. The reason this would be adequate was stated as it being in the best interests of tourism business groups or heritage managers to represent all user positions. However, as has been previously noted in the literature review, seldom are managers able to accurately gauge users' expectations and attitudes (Baker et al., 1997; Turner, 1994; Vining, 1992).

Surveys were considered a valid method to indicate the level of importance of the heritage resource for leisure visitors. This would create a preservation value, similar to other resource use values. Resource sector interviews (industry and ministries) indicated that a quantifiable value about the benefits of non-extraction is required in order to hierarchically place the heritage value of the trail against extraction values. Thus, these stakeholders would use the survey data as a valid method of identifying the importance of the Cariboo Wagon Road as a substantial heritage resource. However, other issues such as buffer zones or permitted levels of harvesting were not considered to be adequately addressed through a survey. Industry and government representatives considered the complexity of issues such as financial compensation for affected stakeholders, the specific demands of resource harvest conditions, and other factors the public is not cognizant of, to limit the applicability of survey results to their own planning needs.

Discussion

This study suggests that heritage leisure visitors may be very receptive to planning processes that solicit their needs and expectations. Two of the primary weaknesses in present resource planning processes that Owen (1998) identifies are the lack of adequate user representation and the large demands for data collection and analysis. Surveys that solicit the needs of potential user groups can improve the availability of planning data that targets leisure QoL indicators and increase the overall representation of stakeholders in shared use land planning. The need for timely and relevant planning data was identified in literature and by interviewees as of crucial importance for all stakeholders. It was felt that non-industrial or governmental stakeholders lacked access to a variety of pertinent planning data and that comprehensive and equitable access to relevant planning data has consistently been a great challenge in shared use land planning processes in this region, at both the regional and sub-regional levels (Penrose et al., 1998).

Respondents to the survey from all resident sub-samples clearly identified the need to solicit the views of all visiting publics (local, national, and international) when making resource allocation decisions. They felt that the participation of visiting publics was considered of higher priority than that of government or industrial resource users, a finding supported by other resource management research (Keogh, 1990; Robinson et al., 1997). This could suggest that personal QoL values (e.g., leisure or free time) have higher importance to some stakeholders than do economic QoL variables. Other results from this study that may have a significant bearing on the LRMP include season and mode of recreational use, conflict between recreation types, heritage impact risks from recreation, and a diversity of industry impact assessments.

Expectancies and Visitor Management

A principal tenet of this study was that different user groups hold differing expectations, would have varied QoL leisure needs, and further that the greater the importance of the activity, the more discerning or critical would be the expectations for quality. However, the survey results do not substantiate this contention and rather suggest a more homogenous set of expectations between various Barkerville visitor groups than would be expected from previous research. Three primary avenues of explanation could account for these near homogenous results: that heritage visitors as a group display similar expectations and characteristics rather than representing diverse populations, that survey responses represent a general environmental aesthetic, or that the expectancy variables as identified in the study failed to elucidate actual expectations.

Recent research into the Canadian public's attitudes toward environmental management has also revealed relatively homogenous results by residency regarding acceptable or unacceptable uses of natural resources (Owen, 1998; Robinson et al., 1997). These results indicate a common QoL environmental value for natural environments, a wish to limit some economic impacts and a preference not to recreate where such activities have occurred. Surprisingly, studies that have occurred in communities dependent upon resource harvesting have still cited this public resistance to consumptive harvest practices or to viewing the effects of such harvests (Forest Development Section, 1997).

Residency does appear to be the largest determinant of impact expectations. Expectancy theory would suggest that distant visitors, who have significantly greater investments in reaching the site, would report higher QoL leisure expectations than local populations. However, the results were contrary to expectation theory, in that those with the greatest investment in reaching the site (international visitors) reported the least opposition to industrial impacts. This should be qualified by stating that all visitor groups still reported negative impacts from motorized recreation or industrial activity on or near the trail. One possible explanation is the influence of whether visitors intended on visiting the site; both USA and international visitors reported the lowest rates of having planned to visit the site before their trip. However, those non-resident visitors who did plan on visiting Barkerville reported longer site visit plans than a local population. Also, both international and USA residents placed the greatest importance on "visiting historical sites to their overall interest in tourism" which should be reflected in increased expectations for a quality experience. This was not supported by the data.

The Heritage Resource and Visitors

Continuing to meet or exceed the satisfaction of the visiting public is both a goal of Wells, the local community dependent upon tourism revenues from visitors, and managers to Barkerville. This becomes part of the rationale for this study and for inclusion of visitors' perspectives into an LRMP system. Over 75% of all survey respondents had indicated that they would likely make use of a heritage trail if one were available. Interviews with heritage staff, local recreational and tourism society representatives indicated that a critical goal of tourism development is to offer a specialized tourism or heritage product to a diversified visitor population and to retain these visitors for a longer period of time. This is supported by previous research done in the Wells–Barkerville area that focused on both the heritage of the region and the expanse of available trails for recreationists as a useful method of encouraging local economic development (Campbell, 1994). Retaining leisure visitors for longer periods with further revenue generation would satisfy local QoL economic variables (e.g., small business development, economic diversification) as well as positively influence visitor QoL leisure outcomes. Moore and Barthlow (1998) highlight the growing popularity of long-distance trails in the USA as a method of attracting and retaining heritage visitors for longer stays.

Heritage tourists have been identified as a significant visitor market, and overseas visitors cite significant interest in visiting heritage, native, and natural sites in BC (Province of BC, 1991). Previous studies at Barkerville (Williams, Tompkins, & Dossa, 1994) have recognized the present limited length of stay for heritage visitors as one of the greatest barriers in the region to increasing tourism revenues. The region presently offers two primary tourist or recreational experiences: canoe excursions or heritage travel, with some crossover traffic within the two activities. Offering additional tourism sites may retain these visitors for longer periods or attract other leisure experience seekers.

Industrial Stakeholders and Visitors

The forest industry generally recognizes that their operations may be compatible with other activities when forestlands are not alienated from industry use. Thus, the forest industry operating in the region often cooperates with local interest groups who have non-commercial mandates (i.e., recreation, tourism, etc.) if their interests are not seen as imperiling timber extraction operations. BC-based forest companies have often modified their operations through the use of partial cutting techniques to meet visual quality objectives and have also relied on horse logging to limit sound or other impacts (Forest Development Section, 1997). Forest industry planning has increasingly focused on conflict mediation and valuing QoL non-economic indicators.

The historical society representative had indicated that the organization presently consults with regional forest firms over harvest plans and activities. Recognizing leisure QoL values of the Cariboo Wagon Road, requests have been made to consider low-scale harvesting near certain portions of the trail. This is to improve viewscapes, where present mature timber restricts the visibility of recreationists. The historical society feels that opening viewscapes in the mature timber could improve recreational aspects of the trail. There is substantial research to suggest that creating viewscapes which reveal mountain peaks, and thinning stands to increase the visual penetration into the stand, will be desirable to many visitors (Forest Development Section, 1997; Hull et al., 1995; Schroeder, 1984). Modifications to the mature forest surrounding the Cariboo Wagon Road are examples of how QoL leisure needs can be accommodated, while maintaining economic values as well.

This study has generated specific trail management concerns around periods of harvesting, location of cutblocks, presence of harvesting equipment, and noise disturbance. Interview data have revealed that forest firms can easily operate in winter, are amenable to horse logging when it is economically viable, are willing to modify harvest procedures when possible, and are required to take historical resources into concern during operations (presently the Wagon Road is not considered a heritage resource by Heritage Conservation Act standards). These conditions suggest that it is possible to incorporate recreational, historical, and economic values together.

Mining interest stakeholders had the greatest resistance to incorporation of non-resident expectations into IRM planning. Survey results clearly indicated that mineral extraction using current technology would result in dissatisfactory experiences for heritage visitors. For placer miners their position in land use planning is clear: They have been granted a Crown right to operate on a specific piece of land and removal of that right is unlawful without just compensation. The limits to cooperation in IRM planning for mining stakeholders may center on the mitigation of avoidable impacts to other parties from their placer operations, and the obeisance of any legal requirements around protection of historical resources. This is an example of how some QoL variables and stakeholders (e.g., leisure and mineral resource extraction) may have irreconcilable conflicts.

Summary

Despite growing interest in expectancy theory in leisure research, the identification of quality and QoL expectations by demographic or investment variables continues to be challenging. This study found that different visiting populations conform to relatively homogenous impact expectations from recreational or industry activities. Respondents largely prefer low-impact, non-mechanized industrial activities for areas experiencing multiple use. The somewhat homogenous results may indicate that heritage visitors display similar environmental expectations or a broader common value of environmental aesthetic preferences. The recent public involvement in managing resource extraction activities suggests that surveys based on expectancy theory could be a tool to quantify leisure QoL values for non-residents.

Heritage tourism is one of the fastest growing forms of tourism worldwide and has been identified as of great interest to BC visitors. Heritage trails set in natural environments are attractive in providing cultural, structural, and natural heritage features simultaneously. Thus, they may satisfy diverse leisure needs, improve rural economic QoL conditions (e.g., economic diversification, small business, employment), and still permit some traditional resource extraction activities undertaken after LRMP programs.

The use of surveys employing expectancy theory is a growing area of recreation and tourism research. Such research is beginning to increase in tourism studies targeting satisfaction levels of visitors (Cho, 1998) and the representation of tourism environments to visitors (Fenton et al., 1998). This study suggests that such uses of expectancy theory require further investigation to be employed in assessing leisure QoL values.

The need to develop more sensitive instruments to detect visitor expectations and satisfaction antecedents is pressing for an industry in which the experience is the main benefit. Recommendations for further research include the examination of expectation development, influence of commitment to QoL value identification, application of audio video and other electronic tools for quantifying effect expectations, and the evolution of expectations by personal investment. This study has provided ample scope for a variety of future research directions.

References

Adams, G. R., & Schvaneveldt, J. D. (1991). *Understanding research methods* (2nd ed). New York: Longman Publishing Group.

Aronsson, L. (1994). Examples of sustainable rural tourism in Sweden. *Journal of Sustainable Tourism, 2*(1–2), 77–92.

Baker, K. G., Hozier, G. C., Jr., & Rogers, R. D. (1994). Marketing research theory and methodology and the tourism industry: A non-technical discussion. *Journal of Travel Research, 32*(3), 03–08.

Baker, D. A., & Fesenmaier, D. R. (1997). Effects of service climate on managers' and employees' rating of visitors' service quality expectations. *Journal of Travel Research, 36*(1), 15–22.

Boniface, P., & Fowler, P. J. (1993). *Heritage and tourism in the global village.* New York: Routledge Publishing.

Bowers, V. (1998). Ghosts of the West Kootenay. *Beautiful BC Magazine, 40*(3), 30–38.

Bright, A. D. (1997). Attitude-strength and support of recreation management strategies. *Journal of Leisure Research, 29*(4), 363–379.

Butlin, R. A. (1993). *Historical geography: Through the gates of space and time.* London: Edward Arnold Publishing.

Campbell, J. A. (1994). The Wells-Goldfields Trail Network: Towards Community Based Tourism in a Very Small Town. *Thesis in partial fulfilment of a master's degree in the faculty of environmental design,* Calgary.

Cho, B. H. (1998). Assessing tourist satisfaction: An exploratory study of korean youth tourists in Australia. *Tourism Recreation Research, 23*(1), 47–54.

de Vaus, D. A. (1991). Surveys in social research (3rd ed). London: UCL Press.

Dillman, D. A. (1978). Mail and telephone survey: The total design method. New York: Wiley.

Downs, A. (1993). Wagon Road north. Surrey, BC: Heritage House Publishing Co.

Duffy, D., Hallgren, L., Parker, Z., Penrose, R., & Roseland, M. (1998). *Improving the shared decision-making model: An evaluation of public participation in Land and Resource Management planning (LRMP) in British Columbia* (Vol. 1). Victoria: Forest Renewal British Columbia.

Duke, C. R., & Persia, M. A. (1994). Foreign and domestic escorted tour expectations of American travelers. In M. Uysal (Ed.), *Global tourist behavior* (pp. 61–77). London: The Haworth Press.

Ewert, A. (1993). Differences in the level of motive importance based on trip outcome, experience level and group type. *Journal of Leisure Research, 25*(4), 335–349.

Ewert, A. (1998). A comparison of urban-proximate and urban-distant wilderness users on selected variables. *Environmental Management, 22*(6), 927–936.

Ewert, A., & Hood, D. (1995). Urban-proximate and urban-distant wilderness: An exploratory comparison between two "types" of wilderness. *Journal of Park and Recreation, 13*(2), 73–85.

Field, B. (1999). Visitor Trail Attribute Expectations for a Proposed Heritage Trail: Potential Impacts for an Integrated Resource Management Regime. *Thesis submitted in partial fulfillment of the requirements for the degree of master of science.* University of Northern British Columbia.

Forest Development Section. (1997). *Visual impacts of partial cutting-summary report: A technical analysis and public perception study.* Victoria: B.C. Ministry of Forests.

Fowler, F. J. (1995). *Improving survey questions: Design and evaluation.* California, CA: Sage Publications, Inc.

Gnoth, J. (1997). Tourism motivation and expectation formation. *Annals of Tourism Research, 24*(2), 283–304.

Go, F. M., Milne, D., & Whittles, L. J. R. (1992). Communities as destinations: A marketing taxonomy for the effective implementation of the tourism action plan. *Journal of Travel Research, 30*(04), 31–37.

Hull IV, R. B., & Stewart, W. P. (1995). The landscape encountered and experienced while hiking. *Environment and Behavior, 27*(3), 404–426.

Institute for Participatory Management and Planning. (1994). *Citizen participation handbook: For public officials and other professionals serving the public* (8th ed). Monterey, CA: IPMP.

Johnson, M. D., Anderson, E. W., & Fornell, C. (1995). Rational and adaptive performance expectations in a customer satisfaction framework. *Journal of Consumer Research, 21*(March), 695–707.

Keogh, B. (1990). Public participation in community tourism planning. *Annals of Tourism Research, 17*, 449–465.

Manfredo, M. J., Driver, R. L., & Brown, P. J. (1983). A test of concepts inherent in experience based setting management for outdoor recreation areas. *Journal of Leisure Research, 15*(03), 263–283.

Ministry of Canadian Heritage. (1994). Guiding principles and operational policies. Ottawa, ON: Ministry of Supply and Services.

Moore, R. L., & Barthlow, K. (1998). *The economic impacts and uses of long distance trails: Featuring a case study of the overmountain victory national historic trail.* Washington, DC: United States Department of the Interior and National Park Service.

Mugica, M., & De Lucio, J. V. (1996). The role of on-site experience on landscape preferences: A case study at Donana National Park (Spain). *Journal of Environmental Management, 47*(July), 229–239.

Neal, J. D., Uysal, M., & Sirgy, J. M. (2007). The effect of tourism services on travelers' quality of life. *Journal of Travel Research, 46*(November), 154–163.

Oliver, R. L. (1980). A cognitive model of the antecedents and consequences of satisfaction decisions. *Journal of Marketing Research, 17*(4), 460–469.

Olson, J. M., Roese, N. J., & Zanna, M. P. (1996). Expectancies. In: E. T. Higgines, & A. W. Kruglanski (Eds.), *Social psychology: Handbook of basic principles* (pp. 211–238). New York: Guildford Press.

Owen, S. (1998). Land use planning in the nineties: CORE lessons. *Environments, 25*(2 & 3), 15–26.

Patenaude, B. C. (1995). *Trails to gold*. Victoria: Horsdal & Schubart Publishers Ltd.

Penrose, R. W., Day, J., & Roseland, M. (1998). Shared decision making in public land planning: An evaluation of the Cariboo-Chilcotin CORE process. *Environments, 25*(2 & 3), 27–47.

Phillips, A., & Tubridy, M. (1994). New supports for heritage tourism in rural Ireland. *Journal of Sustainable Tourism, 2*(1–2), 112–129.

Pigram, J., & Jenkins, J. (1994). *Rural recreation and tourism-policy and planning*. In *D. Mercer (Ed.), New viewpoints in Australian outdoor recreation research and planning* (pp. 119–129). Sydney: Hepper Marriott and Associates Publishers.

Province of BC. (1991). *Overseas visitors to British Columbia*. Ministry of development, trade and tourism. Victoria: Queen's Printer.

Province of BC. (1992). *Heritage planning: A guide for local government*. Victoria: Heritage Conservation Branch.

Province of BC. (1993a). *1993/1994 business plan-draft report*. Victoria: Ministry of Tourism and Ministry Responsible for Culture-Marketing Division.

Purcell, A. T. (1986). Environmental perception and Effect: A schema discrepancy model. *Environment and Behavior, 18*(1), 3–30.

Research Resolutions. (1996). *The northern tourism experience: 1994 domestic and inbound markets to Canada's north*. Toronto, ON: Canadian Tourism Commission.

Robinson, D., Hawley, A., & Robson, M. (1997). *Social Valuation of the McGregor Model Forest Assessing Canadian Public Opinion on Forest Values and Forest Management, Results of the 1996 Canadian Forest Survey*. Published by the McGregor Model Forest Association.

Roszell, J. (1996). Planning and managing natural and cultural resources: A parks Canada perspective. *Environment's: A Journal of Interdisciplinary Studies, 24*(01), 27–41.

Ryan, C. (1995). Researching tourist satisfaction: Issues, concepts, problems. London: Routledge Press.

Schreyer, R., Lime, D. W., & Williams, D. R. (1984). Characterizing the influence of past experience on recreation behavior. *Journal of Leisure Research, 16*(01), 3450.

Schroeder, H. W. (1984). Environmental perception rating scales: A case for simple methods of analysis. *Environment and Behavior, 16*(05), 573–598.

Shelby, B., & Harris, R. (1985). Comparing methods for determining visitor evaluations of ecological impacts: Site visits, photographs, and written descriptions. *Journal of Leisure Research, 17*(01), 57–67.

Singleton, R. A., Jr., Straits, B., & Bruce, M. M. (1993). *Approaches to social research* (2nd ed). New York: Oxford University Press.

Stewart, W. P., & Carpenter, E. H. (1989). Solitude at grand canyon: An application of expectancy theory. *Journal of Leisure Research, 21*(01), 04–17.

Tarrant, M. A., Cordell, K. H., & Kibler, T. L. (1997). Measuring perceived crowding for high-density river recreation: The effects of situational conditions and personal factors. *Leisure Sciences, 19*(02), 97–112.

Turner, A. (1994). Managing Impacts: Measurement and judgement in natural resource management. In D. Mercer (Ed.), New viewpoints in Australian outdoor *recreation research and planning* (pp. 129–140). Sydney: Hepper Marriott and Associates Publishers.

Vining, J. (1992). Environmental emotions and decisions: A comparison of the responses and expectations of forest managers, an environmental group, and the public. *Environment and Behavior, 24*(1), 3–34.

Wade, M. S. (1979). *The Cariboo road*. Victoria: Haunted Bookshop Publishers.

Webb, W. M., & Worchel, S. (1993). Prior experience and expectation in the context of crowding. *Journal of Personality and Social Psychology, 65*(03), 512–521.

Weiler, B. & Hall, Me. (Eds.). (1992). *Special interest tourism.* London: Belhaven Press Ltd.

Williams, P. W., Tompkins, L., & Dossa, K. B. (1994). *BC Heritage Site Visitor Market Profile Study 1994.* Victoria: BC Ministry of Small Business, Tourism, and Culture.

Wilson, A., Roseland, M., & Day, J. E. (1996). Shared decision-making and public land planning: An evaluation of the Vancouver Island regional CORE process. *Environments, 23*(02), 69–86.

Wright, P. A. (1996). North American ecotourists: Market profile and trip characteristics. *Journal of Travel Research, 35*(04), 02–10.

Yale, P. (1991). *From tourist attractions to heritage tourism.* Huntingdon: ELM Publications.

Zeppel, H., & Hall, M. C. (1992). Review, arts and heritage tourism. In Weiler, B., & Hall, M. C. (Eds.), *Special interest tourism* (pp. 47–68). London: Belhaven Press.

Chapter 6
Development of a Tourism and Quality-of-Life Instrument

Kathleen L. Andereck and Gyan Nyaupane

Introduction

There are several reasons for investigating the impact that tourism has on quality of life (QOL) of community residents. One of the most significant implications is to strategically position the tourism industry politically. Currently, economic indicators are available, and the economic impact of tourism as an industry has been presented to legislators and policy makers in many countries, states, and provinces. Yet, often this information does not seem to be coming together with sufficient impact to demonstrate the power of the industry to decision makers, especially as it relates to the needs of the citizenry. Countries have more than an economic stake in tourism though. Each individual is also positively and negatively affected by tourism development in his or her overall quality of living. As an example, the quality and quantity of recreation opportunities in many areas would not exist without a vibrant tourism industry. Nor would a myriad of special event and cultural opportunities exist which contribute to the artistic, educational, and recreational experiences that positively shape the life quality of citizens. Also, the ambiance created in communities by the existence of tourism sets the stage to facilitate the attraction of new businesses and industries. The purpose of this chapter is to present and validate an instrument that measures residents' perceptions of tourism's impact on QOL.

Defining quality of life (QOL) is difficult because it is a subjective experience dependent on individuals' perceptions and feelings. There are over 100 definitions and models of QOL, though there is agreement in recent years that it is a multidimensional and interactive construct encompassing many aspects of people's lives and environments (Schalock & Siperstein, 1996). Quality of life refers to one's satisfaction with life and feelings of contentment or fulfillment with one's experience in the world. It is how people view, or what they feel about, their lives. Similar situations and circumstances may be perceived differently by different people. Therefore,

K.L. Andereck (✉)
Arizona State University, Phoenix, AZ, USA
e-mail: Kathleen.Andereck@asu.edu

M. Budruk, R. Phillips (eds.), *Quality-of-Life Community Indicators for Parks, Recreation and Tourism Management*, Social Indicators Research Series 43, DOI 10.1007/978-90-481-9861-0_6, © Springer Science+Business Media B.V. 2011

many argue QOL is best studied from the perspective of the individual (Taylor & Bogdan, 1990).

With the recognition that tourism has great potential to affect the lives of community residents, over the past several years a number of studies have considered residents' attitudes toward tourism and the impacts tourism can have on a community (Andereck & Vogt, 2000; Gursoy & Rutherford, 2004; Jurowski, Uysal, & Williams, 1997; Ko & Stewart, 2002; Lankford & Howard, 1994; McGehee & Andereck, 2004; Perdue, Long, & Allen, 1987; Teye, Sirakaya, & Sonmez, 2002; Wang & Pfister, 2008). While implicit in this research is the tenet that tourism influences quality of life (QOL) in a community, few studies have developed instruments to directly investigate residents' perceptions of the impact tourism has on their QOL. The difference between quality of life and attitudes/impacts studies is essentially one of the measurement: attitude/impact studies largely focus on the way in which residents feel tourism affects *communities* and the *environment*, whereas quality of life studies are typically concerned with the way these impacts effect individual or family *life satisfaction*, including satisfaction with community, neighborhood, and personal satisfaction (Allen, 1990). Attitude and impact studies are concerned with community changes and support for tourism development. There is an assumed connection between community characteristics and life satisfaction. Attitude studies, however, have generally asked residents to agree or disagree with statements regarding tourism's perceived impacts on their community without specific questions linking these impacts with influences on individuals' life satisfaction or quality of life.

Few tourism studies have measured QOL in the ways it is most often measured in sociological or psychological studies. Most often researchers work to develop social indicators that can be used to describe, predict, and improve quality of life (Massam, 2002). To measure QOL, generally two approaches have been used: (1) indicators that consider objective circumstances of people's lives such as income and education attainment and (2) indicators that consider subjective evaluation of life circumstances, such as satisfaction with various aspects of life (Heal & Sigelman, 1996). Though there is disagreement in the literature regarding the utility of the two kinds of measures, some researchers suggest that QOL indicators should include respondents' assessment of their lives because without such measures results do not capture people's life experiences (Forward, 2003; Rogerson, 1999). Rogerson (1999) proposed that QOL involves identification of individuals' preferences and an evaluation of those preferences. Measures can also be absolute or relative, indexing people's QOL or comparing it to some standard such as what they would ideally want (Heal & Sigelman, 1996). Studies can also measure general aspects of QOL or specific aspects such as community services and how these relate to satisfaction with the community. As well, the unit of analysis for QOL studies can range from the individual to the world with the individual, family, or community being common units of analysis (Rogerson, 1999; Sirgy, Rahtz, Cicic, & Underwood, 2000).

There are myriad instruments that have been developed to measure QOL. Nearly all quantitative subjective QOL indicators are measured with rating scales, almost always assessing satisfaction of various life domains using a series of Likert-type scales or semantic differential scales. They encompass numerous types of indicators from the very general, such as perceptions of health, to the specific, such as loss of wildlife habitat (Forward, 2003). More recently, researchers have made the case that measuring only satisfaction is not adequate in that it assumes equal weighting among the domains (Felce & Perry, 1995). It has been suggested that an overall appraisal of QOL across separate domains should take account of the importance an individual places on the domain (Cummins, 1992; Ferrans & Powers, 1985). It is only possible to interpret an individual's assessment of domains within the context of the importance an individual places on the domain (Felce & Perry, 1995). Edgerton (1990) noted that only an individual can determine trade-offs between the various aspects of his or her own personal welfare. Toward this end, a number of measurement instruments have been developed that measure both respondents' perceptions of importance of indicators as well as assessment of satisfaction with indicators (Brown, Raphael, & Renwick, 1998; Cummins, 1992; Felce & Perry, 1995; Ferrans & Powers, 1985; Massam, 2002). These two measures are then used to create a weighting structure (Felce & Perry, 1995; Forward, 2003). In this way, a QOL score reflects more than satisfaction; it also reflects the extent to which an individual values a domain (Edgerton, 1990; Ferrans & Powers, 1985). (Felce & Perry, 1995) concluded that an ideal model of QOL integrates subjective and objective indicators, encompasses a broad range of domains, and includes a rating of importance of each domain.

In tourism, one study that used an importance–satisfaction type of QOL measure was conducted by Allen, Long, Perdue, & Keiselbach (1988). They measured several dimensions of community life and determined that level of tourism development affects residents' evaluation of community life dimensions. Their indicators, however, were not specific to tourism. Studies that have included tourism impact items specifically have most often used agreement scales to determine residents' perceptions of tourism's effect on a community or their support for tourism development (Andereck & Vogt, 2000; Ap & Crompton, 1998; Choi & Sirakaya, 2005; Lankford & Howard, 1994; Perdue, Long, & Allen, 1990; Wang & Pfister, 2008). Three researchers have developed and validated instruments using these kinds of agreement scales. Lankford and Howard (1994) developed an instrument called the Tourism Impact Attitude Scale (TIAS). The instrument was pretested, refined, and retested. The final version of the TIAS included 27 items measured on five-point agreement scales. Exploratory factor analysis resulted in two dimensions: one with a tourism promotion, impacts, and planning focus and the other with a community development and personal benefits focus. Ap and Crompton (1998) also developed an attitudes instrument using a pretest and purification procedure, with the final version containing 35 items with five-point tourism decreases or increases scales and five-point dislike to like scales. The two sets of measures were used to calculate an index. Exploratory factor analysis resulted in seven domains: social and cultural, economic, crowding and congestion, environmental, services, taxes, and community attitude. Finally, Choi and Sirakaya (2005) developed the SUS-TAS, an instrument with a sustainability focus. The instrument was pretested, refined, and retested resulting in a final instrument that included 48 items measured on a five-point agreement scales. Seven reliable domains were discovered using exploratory factor analysis: perceived social costs, environmental sustainability, long-term planning, perceived economic benefits, community-centered economy, ensuring visitor satisfaction, and maximizing community participation.

Policy makers need information that demonstrates how an area is doing that incorporates how citizens perceive the factors that contribute to their own quality of life. To that end it is important to examine the perspective of community residents in relation to how they experience tourism; in other words, the extent to which residents feel tourism influences aspects of community life that they deem as personally valuable. Toward that end, we propose a tourism and quality of life (TQOL) instrument that more completely measures the perceived implications of tourism for quality of life. We combine one of the common methods for QOL assessment, that is an assessment of the importance of tourism-related indicators and assessment of satisfaction with those indicators similar to Massam (2002), with an assessment of the way residents feel tourism affects those indicators.

Instrument Development

Focus Groups

To begin the process of developing the measurement instrument after an extensive literature review about tourism and resident attitudes toward tourism, several focus groups were held with tourism professionals. The goal of the focus group experiences was to generate discussion on the dimensions of quality of life influences of tourism. A focus group consists of people with a common characteristic who provide qualitative data through focused discussion. This method stimulates new ideas among participants, who can direct their thoughts both toward one another and the interviewer. Focus groups are often used for exploratory research. Focus groups are also an effective way to establish questions to be asked in a future questionnaire (Krueger, 1991). The major limitation of the focus group method is the potential lack of generalizability to a broader population. In the case of this study, for example, the discussion results may or may not richly represent the views of all tourism professionals in Arizona or all citizens in Arizona. Yet, the power of focus groups is that great detail is generated and there is tremendous potential for gaining insight on the dimensions of the topic at hand. Thus, the results presented here provide substantial insight into the quality of life effects of tourism as perceived by residents. It is important to remember, however, that they reflect only the opinions and perceptions of the group participants and cannot be assumed to represent professionals in the tourism industry as a whole or the general populace of Arizona. To accomplish this, seven focus group interviews were conducted with tourism professionals representing communities and interests throughout the state.

To maximize convenience for participants, the focus groups were convened in different locations scattered across the state. Participants were selected through purposive sampling based on recommendations from members of the Arizona Tourism Alliance Research Committee and other tourism professionals. Each focus group consisted of the ideal four to nine members with representatives from numerous communities and segments of the tourism industry. A total of 45 people participated in one of the seven focus group sessions. Procedures for devising the focus groups followed accepted standards in the social sciences (Krueger, 1991):

(1) Initial contact was made with prospective participants as recommended by tourism professionals. Following a loose telephone script, individuals were briefed on the objectives of the study and asked to participate.
(2) To those who agreed to participate, a confirmation letter stating the time, date, and location of the meeting was faxed within a few days of the initial contact.
(3) A set of three exercises were used to trigger discussion during the meetings, which lasted approximately 2–2.5 h.
(4) After all seven meetings, personalized "thank you" notes were mailed to each participant.

The discussion-guiding questions were oriented toward the way tourism affects participants' personal quality of life both positively and negatively, as well as the way tourism might affect the quality of life of others. To gather background data on the participants, a brief questionnaire was given to participants. Following a warm-up to establish group comfort and rapport, one facilitator led the groups through three types of exercises. First, participants were asked to discuss the ways tourism affects the quality of their lives positively and negatively, as well as to consider economic, socio-cultural, environmental, and political implications of tourism. Participants were then asked to address the same idea through the perspective of different types of people, including a city council member, a small business owner, a corporate CEO, an artist, a parent, and finally as themselves from a professional perspective and as a member of their community. Finally, participants were asked to envision an idyllic community. They were then asked to specify which elements of their idyllic community were the most important, and how tourism might contribute to attaining that idyllic state.

Detailed notes of focus group interviews were taken by a second facilitator, while a third facilitator outlined general concepts on a flip chart. Additional written notes and worksheets by participants and the facilitator(s) were collected. Data analysis followed standard focus group practice (Krueger, 1991).

(1) Following each focus group, all of the flip chart pages, notes, and worksheets were synthesized into one document.
(2) After all the sessions and the worksheets were transcribed, all the available information was thoroughly read.
(3) The information from each meeting was read again with concentration on separate issues by considering the words; considering the context; considering the

internal consistency; considering the specificity of responses; finding the big ideas; and considering the social effect between participants.

(4) Throughout the process, the objectives of the focus groups were considered.

(5) Standardized coding processes were followed (Strauss & Corbin, 1990), including the use of open coding (breaking the data into larger categories) and axial coding creating new categories by identifying relationships between the initial categories.

Table 6.1 captures the range of issues that were introduced by the focus group conversations with respect to the challenges offered by tourism to quality of life. Four categories of impacts are presented: socio-cultural, economic, environmental, and political. Within each category, the emergent themes are presented in order of the magnitude of interest. Themes most frequently mentioned and discussed are presented first and themes that were less frequently mentioned and discussed are presented last. Subthemes within the broader issue are included in certain cases.

The positive contributions of tourism to the quality of life of Arizona residents are many. Focus groups participants were asked to articulate the positive impacts of tourism that they and others perceive within their communities. Table 6.2 summarizes myriad positive impacts that were generated by the seven focus groups. As was done for the discussion on tourism challenges above, four categories of impacts are presented: economic, socio-cultural, environmental, and political. Within each category, the emerging themes are presented in order of the magnitude of interest and discussion that was generated across the groups. Themes most frequently mentioned and discussed are presented first and themes that were less frequently mentioned and discussed are presented last.

Questionnaire Construction

To measure residents' perceptions of tourism impacts on QOL, a questionnaire was designed to measure 38 indicators using existing QOL types of measures in combination with resident attitudes toward tourism studies including (Andereck & Vogt, 2000; Brown et al., 1998; Lankford & Howard, 1994; Massam, 2002; McCool & Martin, 1994; Perdue et al., 1990) and the focus group results. The first group of items was developed to measure the importance respondents placed on the indicators on a five-point Likert-type scale that ranged from 1 ("not at all important") to 5 ("extremely important"). The second set of measures considered the same 38 tourism-related quality of life indicators and used a five-point Likert-type scale that ranged from 1 ("not at all satisfied") to 5 ("extremely satisfied"). The final set of measures was based on resident attitudes toward tourism research and measured perceptions of the effects of tourism on the 38 indicators with minor wording changes (denoted parenthetically in Table 6.3) using a five-point Likert-type scale where 1 = "tourism greatly decreases," 2 = "tourism decreases," 3 = "tourism has no effect," 4 = "tourism increases," and 5 = "tourism greatly increases." Table 6.3 summarized the means for these three sets of scales.

Table 6.1 Challenges to the enhancement of quality of life through tourism

Socio-cultural challenges	Economic challenges
Friction/conflict between residents and tourists	Increased cost of living: real estate/housing/property/rent, prices of goods and services
Lack of understanding, respect for local people, especially Native Americans cultural clash	Big businesses vs. local small businesses
Lack of authenticity: "fake" art, loss of integrity	Stress upon inadequate infrastructure
Exploitation of artists	Seasonality
Trivialized/contrived/watered down culture, heritage, history	Seasonal economy
Disrupt way of life	Seasonal jobs
Less privacy	Vacant/substandard housing and buildings during low season
Loss of sense of place	Unskilled/uncommitted labor
Resident resistance to change/growth	Low-end/low-paying jobs
Antitourism sentiments	Competition from other destinations
Crowding and congestion	Stress on services and staff
Lack of safety	Interruption of daily economic activity
Strangers	Overreliance on tourism industry
Traffic	Lack of tax revenue going to tourism
Unsafe behavior	Fragile nature of industry/economy dependence
Crime	Price gouging
Long work hours	Economic separation between tourists and residents
Negative view of tourists by residents	Oversupply (of hotels, restaurants, etc.)
Drug and alcohol abuse	Short-term investment
Transient populations	
Vandalism	
Noise	

Environmental challenges	Political challenges
Growth: overdevelopment, population, visitation; urbanization/erosion of small town flavor; sprawl; access problems	Factions/conflicts between residents and political entities
Traffic and congestion	Competition for and disagreement over resource distribution
Damage and destruction of natural resources impacts on natural ecosystems	State-level promotional challenges: limitations and conditions of TIFS
Litter	Funding and services prioritize tourists over residents
Wildlife and plant habitat destruction	Policy-maker conflict over community interests vs. tourism interests
Air and water pollution	
Parking problems	
Historic/cultural site destruction	
Aesthetic impact	

Table 6.2 Strengths of the tourism industry in enhancing quality of life

Socio-cultural strengths	Economic strengths
Builds awareness/appreciation of cultural heritage	More businesses, greater diversity of businesses
Exposure to culture and heritage	Increased income
Preservation of culture and heritage	Increased tax revenue
Cultural understanding and knowledge	Tax revenue
Increased global/cultural awareness and understanding	Lower taxes for residents due to tax revenue
Appreciation for diversity	Funds for services – city projects, education
Expanded world-view	Increased employment
Increased special events	Entry level jobs for teens
Cultural and arts events	More customers, demand for products and services
Family events, activities for children	Development of nontourism businesses
Sports	Increased entrepreneurial migration
Greater recreation opportunities and amenities	Improved business climate
Outdoor recreation, trails	Enhanced business creativity
Golf	Increased investment
Children's activities	Increased property values
Increased quality and quality of retail stores, restaurants, accommodations	Greater economic independence
Attractions	Emergence of new communities
Museums	Cooperative promotion of community
Cultural and arts attractions	
Increased residential in-migration	
Increased diversity of cultures and talents	
Increased workforce and volunteers	
Heightened awareness of community, state	
Improved image	
Awareness of tribes	
Provides rewarding field for tourism professionals	
Increased educational opportunities	
Cultural programs/attractions	
Increased community pride	
Expanded social services	
Better infrastructure	
Roads, airport, sewer, water	
Greater opportunities to build relationships	
Enhanced community solidarity	
Increased sense of making a difference	
Improved customer service	
Maintain desired community atmosphere	
Increased volunteer base (winter visitors)	
Expanded children's career horizons	
Safe community	

Table 6.2 (continued)

Environmental strengths	Political strengths
Revitalization/preservation of historic/cultural buildings and sites	Increased partnerships, cooperation, collaboration, regionalism
Provides beautification of community	Public/private/tribal entities
Promotes a clean environment	Enhanced image of tourism among policy makers
Increases natural preserves	Increased debate and dialog about tourism
Increased environmental awareness and appreciation	Opportunity to voice opinion
Community commitment to environment	Political self-destiny
Improved environmental quality and protection	

Table 6.3 Means for QOL indicators

Items	Importance[a]	Satisfaction[b]	Tourism effects[c]
Preserving (peace and quiet)	4.48	3.23	0.38
Feeling safe	4.66	3.41	1.09
Clean air and water	4.75	3.04	0.60
City services like police and fire protection	4.56	3.77	1.46
A stable political environment	4.01	2.99	1.15
Good public transportation	3.60	2.41	1.50
The beauty of my community	4.36	3.42	1.34
Quality of roads, bridges, and utility services	4.41	3.09	1.45
The prevention of (crowding and congestion)	4.41	2.72	−1.01
Controlled (traffic)	4.42	2.70	−1.23
Controlled (urban sprawl and population growth)	4.34	2.56	−0.48
(Litter) control	4.47	2.93	−0.46
Proper (zoning/land use)	4.28	2.85	0.00
My personal life quality	4.66	3.83	0.90
The preservation of my way of life	4.32	3.42	0.72
A feeling of belonging in my community	3.99	3.37	1.05
A stable political environment	3.79	3.07	1.05
Having tourists who respect my way of life	3.86	3.23	0.95
The image of my community to others	3.98	3.43	1.79

Table 6.3 (continued)

Items	Importance[a]	Satisfaction[b]	Tourism effects[c]
An understanding of different cultures	3.81	3.22	1.40
Awareness of natural and cultural heritage	3.74	3.36	1.69
Community pride	4.12	3.36	1.50
Opportunities to participate in local culture	3.36	3.30	1.33
Preservation of wildlife habitats	4.22	3.09	1.15
Preservation of natural areas	4.35	3.11	1.28
Preservation of cultural/historical sites	4.16	3.24	1.57
Strong and diverse economy	4.12	3.15	1.75
Stores and restaurants owned by local residents	3.58	3.13	1.26
The value of my house and/or land	4.41	3.67	1.36
Enough good jobs for residents	4.28	2.92	1.52
Plenty of retail shops and restaurants	3.62	3.49	2.02
Fair prices for goods and services	4.28	3.20	1.07
Plenty of festivals, fairs, museums	3.48	3.28	2.02
Having live sports to watch in my community	3.17	3.34	1.81
Quality recreation opportunities	4.10	3.30	1.67
The prevention of crime and vandalism	4.71	3.13	-0.30
The prevention of drug and alcohol abuse	4.54	2.91	0.16
Tax revenue (sales tax/bed tax)	3.89	2.97	1.61

[a]Scale: 1 = not at all important to 5 = extremely important.
[b]Scale: 1 = not at all satisfied to 5 = extremely satisfied.
[c]Scale: for positive items –3 tourism greatly decreases to +3 tourism greatly increases, for negative items +3 tourism greatly decreases to –3 greatly increases; revised wording in parentheses.

The data were collected from a random sample of Arizona adult residents, proportionally stratified by county. A state-wide telephone survey was conducted by a telephone survey company to elicit study participation, gather names and addresses, and ensure county proportions. Of those contacted by phone, 35% met the survey qualifications (Arizona resident, county of residence, 18 years old or over) and agreed to participate in a survey. A mail-back questionnaire, cover letter, and postage-paid envelope were then sent to 1,222 qualified residents. As suggested by Dillman (2000), a postcard reminder followed the questionnaire by one week, with a second questionnaire, cover letter, and postage-paid envelope mailed to nonrespondents two weeks after the postcard. A map of Arizona public recreation sites and entry in a drawing for an "ArizonaGear" gift set were used as incentives.

Controlling for bad addresses (230), response rate for the mail survey was 70% resulting in a sample of 695.

Testing of the Instrument

State-Wide Sample

Of the respondents, 55% were female; the mean age was 54; and 79% were Anglo, 19% were Hispanic/Latino(a), and 4% of the respondents represented other ethnic or racial backgrounds. Most respondents reported household incomes from $20,000 to $59,999 (54%) and education levels beyond a high school diploma (80%). A comparison of respondents with Arizona's general population revealed that the rural to urban proportion mirrored the population. As is often the case with survey research, however, the sample had a higher average age, education, and income than state residents in general. The Hispanic proportion was representative; however, other ethnic minorities were underrepresented, especially Native Americans. These biases should be kept in mind with respect to study results.

Developing Tourism and Quality of Life Constructs

A tourism and quality of life measure was developed using a series of calculations. An index developed by Brown et al. (1998) and modified by Massam (2002) was used to develop a QOL measure. Their method uses importance and satisfaction ratings of items to determine a QOL score ranging from –10 to +10. For example, an item rated as extremely important with which a respondent is extremely satisfied receives a score of +10. If the item is extremely important and the respondent is not at all satisfied an item is given a score of –10. Items then range between the two end points depending on the importance and satisfaction ratings (see Brown et al., 1998; Massam, 2002 for more details). For the purpose of calculating a quality of life score, the measures have been modified further so they range from 1 to 20 without any zeros and negative scores to facilitate ease of calculations (see Table 6.4).

A tourism and quality of life index was then computed by using the respondents' perceptions of tourism's effect on QOL in conjunction with the 1–20 QOL indicators calculation. First, the items (1–5 point scale) were recoded into scores ranging from –3 to +3 where $1 = -3, 2 = -2, 3 = 1, 4 = 2$, and $5 = 3$. Six negative statements, such as "tourism increases crime," were recoded in reverse order. The perceptions scores were then multiplied by the QOL scores. For example, an item with a QOL score of 20 (very important and very satisfied) and a perceptions rating of +3 (tourism greatly increases) results in a tourism and quality of life (TQOL) score of +60; if the perceptions rating was a –3 (tourism greatly decreases) however, the TQOL score is a –60. Thus, the TQOL score not only represents the extent to

Table 6.4 Calculation of QOL scores using importance and satisfaction

Importance	Satisfaction	Brown et al.'s QOL	New quality of life score
5	5	+10	20
	4	+5	15
	3	0	10
	2	−5	5
	1	−10	1
4	5	+8	18
	4	+4	14
	3	0	10
	2	−4	6
	1	−8	2
3	5	+6	16
	4	+3	13
	3	0	10
	2	−3	7
	1	−6	4
2	5	+4	14
	4	+2	12
	3	0	10
	2	−2	8
	1	−4	6
1	5	+2	12
	4	+1	11
	3	0	10
	2	−1	9
	1	−2	8

Modified from Brown et al., 1998, p. 16.

which tourism is perceived to influence a QOL indicator, it also denotes an individual's value judgment of the indicator by including a measure that incorporates both importance of and satisfaction with that indicator. Negative scores denote tourism is playing a negative role for quality of the life (Table 6.5).

Next, tourism and quality of life domains (TQOL) were developed using exploratory factor analysis (EFA). A sample of 347 participants was randomly selected using a randomized split method. Principal component factor analysis with varimax rotation of TQOL items resulted in seven factors with items that loaded reasonably well and have fairly strong reliability (Table 6.6). Although one domain, tax and amenities, has a marginally acceptable alpha coefficient (0.55), given the exploratory nature of this research and that the domain makes conceptual sense, it was retained. Only those items that have a factor loading of 0.4 or above were included in the factor. A total of nine items were excluded after the factor analysis as these items either were double loaded (0.4 or above) on two factors or were not conceptually tied with the factor (Thurstone, 1947). The seven domains developed include (1) *personal and community life* which includes four items related to community pride and individuals' way of life; (2) *community wellbeing* includes four

Table 6.5 Means for QOL indicators

Items	QOL score[a]	TQOL score[b]
Preserving (peace and quiet)	11.07	4.79
Feeling safe	12.03	12.76
Clean air and water	10.22	7.39
City services like police and fire protection	13.72	20.09
A stable political environment	10.06	12.06
Good public transportation	7.74	12.11
The beauty of my community	11.95	16.27
Quality of roads, bridges, and utility services	10.46	16.24
The prevention of (crowding and congestion)	8.75	−8.62
Controlled (traffic)	8.65	−11.30
Controlled (urban sprawl and population growth)	8.01	−3.98
(Litter) control	9.79	−5.56
Proper (zoning/land use)	9.37	0.05
My personal life quality	14.00	13.07
The preservation of my way of life	11.97	9.21
A feeling of belonging in my community	11.77	13.41
A stable political environment	10.33	11.12
Having tourists who respect my way of life	10.97	11.72
The image of my community to others	11.81	21.59
An understanding of different cultures	10.95	15.88
Awareness of natural and cultural heritage	11.51	19.88
Community pride	11.66	17.86
Opportunities to participate in local culture	11.20	15.33
Preservation of wildlife habitats	10.42	13.19
Preservation of natural areas	10.53	14.98
Preservation of cultural/historical sites	11.06	18.44
Strong and diverse economy	10.74	19.27
Stores and restaurants owned by local residents	10.51	13.97
The value of my house and/or land	13.11	18.36
Enough good jobs for residents	9.71	15.83
Plenty of retail shops and restaurants	11.88	24.44
Fair prices for good and services	10.95	11.94
Plenty of festivals, fairs, museums	11.11	22.86
Having live sports to watch in my community	11.40	21.16
Quality recreation opportunities	11.31	19.53
The prevention of crime and vandalism	10.66	−4.04
The prevention of drug and alcohol abuse	9.65	1.46
Tax revenue (sales tax/bed tax)	10.00	16.81

[a]Range: 1–20.
[b]TQOL score = QOL × tourism effects, range: −60 to 60.

Table 6.6 Factor analysis of TQOL domains

Domains	Factor loadings	Eigen-value	Variance explained (%)	Alpha
Personal and community life (TQOLPC)		9.84	25.89	0.76
Community pride	0.722			
A feeling of belonging in my community	0.711			
The preservation of my way of life	0.632			
My personal life quality	0.630			
Community well-being (TQOLWELL)		3.7	9.73	0.75
Preserving peace and quiet	0.733			
Clean air and water	0.727			
Feeling safe	0.644			
City services like police and fire protection	0.587			
Natural/cultural preservation (TQOLPRES)		2.0	5.23	0.88
Preservation of wildlife habitats	0.836			
Preservation of natural areas	0.817			
Preservation of cultural/historical sites	0.664			
Urban issues (TQOLURBAN)		1.72	4.53	0.77
Controlled traffic	0.787			
Controlled urban sprawl and population growth	0.768			
The prevention of crowding and congestion	0.731			
Litter control	0.586			
Proper zoning/land use	0.507			
Tax and amenities (TQOAMENITIES)		1.45	3.81	0.55
Tax revenue (sales tax/bed tax)	0.621			
Stores and restaurants owned by local residents	0.527			
Plenty of festivals, fairs, museums	0.506			
Economic strength (TQOLECON)		1.40	3.69	0.66
The value of my house and/or land	0.665			
Strong and diverse economy	0.654			
Enough good jobs for residents	0.632			
Crime and substance abuse (TQOLCRIME)		1.13	3.00	0.70
The prevention of drug and alcohol abuse	0.687			
The prevention of crime and vandalism	0.572			
Excluded variables:				
Fair prices of good and services				
A stable political environment				
Good public transportation				
Quality of roads, bridges, and utility services				
Resident participation in local government				
Having tourists who respect my way of life				

Table 6.6 (continued)

Domains	Factor loadings	Eigen-value	Variance explained (%)	Alpha
Awareness of natural and cultural heritage				
Quality recreation opportunities				
Having live sports to watch in my community				
Kaiser–Meyer–Olkin (KMO) = 0.875; KMO varies from 0 to 1 and it should be 0.6 or higher to proceed with factor analysis				
Barlett's test of spheriosity 5,596, df = 703, p < 0.001				

items related to peace, safe, and clean community; (3) *natural/cultural preservation* which includes the three preservation-oriented items; (4) *urban issues* which includes five items typically considered negative impacts of tourism and often associated with urban areas; (5) *tax and amenities* consisting of three items related to tax revenue, stores and restaurants, and festivals; (6) *economic strength* which includes three items related to economic impacts; and (7) *crime and substance abuse* that is made up of the two crime-oriented items.

A confirmatory factor analysis (CFA) was conducted using AMOS 17.0 to refine and confirm the exploratory domains developed. The remaining 348 participants were used for the CFA analysis. An examination of the fit of the measurement model with seven dimensions with 24 items revealed good fit with the data (Table 6.7). However, modification indices suggested that two of the items should be deleted because these items had poor factor loadings (<0.4) and had very low corresponding R^2 values (0.11 and 0.12). The R^2 values suggested that their contributions to the variance explained were not significant. These two items included city services like police and fire protection and the value of my house and/or land. In order to improve the model, these two items were deleted. After deleting the two items, the fit of the model was significantly improved.

CFA models were evaluated using the χ^2 goodness-of-fit test and other fit indexes (Table 6.8). The χ^2 goodness-of-fit statistic assesses the magnitude of discrepancy between the sample and fitted covariance matrices. The smaller the value, the better the model. Since χ^2 is very sensitive to sample size, it might not be the best indicator by itself (Hu & Bentler, 1999). If the sample is large enough, a desirable χ^2 value is less than 3 (Kline, 1998). Fit indexes, which are supplemental to the χ^2 test, can be classified into absolute and incremental fit indexes. An absolute fit index evaluates how well an a priori model reproduces the sample data. Goodness-of-fit index (GFI), adjusted goodness-of-fit (AGFI), standardized root mean squared residual (SRMR), and the root mean square error of approximation (RMSEA) are examples of absolute fit indexes (Hu & Bentler, 1999). Both GFI and AGFI can be out of the range of 0–1

Table 6.7 Results of the confirmatory factor analysis for the quality of life domains

Tourism quality of life items	Estimate (B)	R^2
Personal and community life (TQOLPC)		
Community pride	0.79	0.63
A feeling of belonging in my community	0.78	0.61
The preservation of my way of life	0.64	0.41
My personal life quality	0.40	0.16
Community wellbeing (TQOLWELL)		
Preserving peace and quiet	0.80	0.73
Clean air and water	0.77	0.60
Feeling safe	0.69	0.48
Natural/cultural preservation (TQOLPRES)		
Preservation of wildlife habitats	0.86	0.72
Preservation of natural areas	0.97	0.93
Preservation of cultural/historical sites	0.74	0.55
Urban issues (TQOLURBAN)		
Controlled traffic	0.72	0.52
Controlled urban sprawl and population growth	0.58	0.33
The prevention of crowding and congestion	0.76	0.58
Litter control	0.64	0.41
Proper zoning/land use	0.48	0.23
Tax and amenities (TQOLAMENITIES)		
Tax revenue (sales tax/bed tax)	0.40	0.16
Stores and restaurants owned by local residents	0.51	0.26
Plenty of festivals, fairs, museums	0.52	0.27
Economic strength (TQOLECON)		
Strong and diverse economy	0.69	0.48
Enough good jobs for residents	0.67	0.45
Crime and substance abuse (TQOLCRIME)		
The prevention of drug and alcohol abuse	0.72	0.51
The prevention of crime and vandalism	0.68	0.46

Table 6.8 Summary of nested models of tourism quality life domains

Model	Scaled χ^2	df	GFI	CFI	NFI	RMSEA	$\chi^2 \Delta$ test
Initial model (based on EFA)	439.67	231	0.900	0.915	0.839	0.053	
Modified models – 2 items deleted	355.73	188	0.911	0.929	0.862	0.053	83.49***

***$p < 0.001$.

and the desirable value is greater than 0.90. Acceptable range of fit for SRMR is less than 0.08 and RMSEA is less than 0.06. Examples of incremental fit indexes include the Normed Fit Index (NFI), Comparative Fit Index (CFI), and Relative Noncentrality Index (RNI). It is desirable to have incremental fit indexes more than 0.90 (Kline, 1998).

Conclusion

The purpose of this study was to test the reliability of a tourism and quality of life measurement instrument. The results of this study should be considered preliminary in that this is the first attempt to validate the instrument. Added research that uses the instrument in its entirety, in the refined version, or adds to the indicators will allow further development of the TQOL indicators. The instrument combines importance and satisfaction measures often used in QOL research with traditional resident attitudes research. The instrument allows for the inclusion of perceptions of life satisfaction which is conceptually the goal of QOL studies. This combination of measures allows researchers to assess residents' preferences and evaluation of those preferences as suggested by Rogerson (1999) and in keeping with many QOL studies (Brown et al., 1998; Cummins, 1992; Felce & Perry, 1995; Ferrans & Powers, 1985; Massam, 2002). This not only provides a measure of satisfaction with an indicator but also allows residents to place an individual value on the indicator (Felce & Perry, 1995; Ferrans & Powers, 1985). In addition, resident attitude types of measures have been incorporated into the instrument to capture an additional component: the manner in which residents perceive tourism impacts their communities. The result is an instrument with a number of indicators that can be used to calculate a tourism and quality of life index.

The newly developed TQOL index has the potential to be an effective and valuable instrument for monitoring residents' tourism experiences. It identifies what is important to residents and how residents interpret the effect of tourism experiences on their own lives. Consequently, the TQOL index is useful for measuring the subjective nature of quality of life. When the TQOL index is used in conjunction with objective and external indicators, a more accurate description of the impact of tourism on the quality of life of residents can be created. An added benefit of the index is its ability to identify inequalities in the distribution of the costs and benefits of tourism to various segments within the community. Thus, the index responds to the recommendations of economists that advocate noneconomic measures to justify the benefits and the costs of growth. The TQOL index can be effective in determining when increased tourism is reducing rather than improving the quality of life (Massam, 2002).

Some aspects of the factor analysis with the TQOL are similar to that of traditional resident attitudes research, while other aspects differ. As compared to resident attitude studies, the domains found in this analysis are somewhat different. There are positive and negative impacts types of factors similar to Andereck and Vogt (2000), Perdue et al. (1990), Teye et al. (2002), and Wang and Pfister (2008). There are also quality-of-life related factors consistent with some resident attitude studies (Andereck & Vogt, 2000; Liu, Sheldon, & Var, 1987). These domains allow for an improved understanding of the way in which residents perceive tourism influences their QOL. They do perceive that tourism has a positive influence on their QOL; they perceive that tourism positively influences the economy, facilitates preservation of natural and cultural resources, can enhance community well-being, and has an overall positive influence on their way of life. On the other hand, residents also recognize that tourism can have negative QOL consequences such as more crime and urban issues, though these are not perceived as highly problematic.

These results suggest residents' quality of life is positively and negatively affected by tourism. Although educating residents about the economic benefits of tourism is important and may result in awareness of tourism as a community economic development strategy, it is clear that there are other community quality of life attributes that are as important, or more so, to local residents. Tourism planners and industry leaders must keep this in mind when developing tourism and trying to gain support for the industry. As well, there is a widespread belief in the tourism industry that the "general public" does not understand or recognize the positive contribution tourism makes to community life. This study also tends to belie this commonly held belief as residents do seem to recognize both the negative and the positive influences of tourism.

References

Allen, L. R. (1990). Benefits of leisure attributes to community satisfaction. *Journal of Leisure Research, 22*, 183–196.

Allen, L. R., Long, P. T., Perdue, R. R., & Keiselbach, S. (1988). The impact of tourism development on residents' perceptions of community life. *Journal of Travel Research, 27*(1), 16–21.

Andereck, K. L., & Vogt, C. A. (2000). The relationship between residents' attitudes toward tourism and tourism development options. *Journal of Travel Research, 39*(1), 27–36.

Ap, J., & Crompton, J. L. (1998). Developing and testing a tourism impact scale. *Journal of Travel Research, 37*(2), 120–130.

Brown, I., Raphael, D., & Renwick, R. (1998). *Quality of life profile (item #2)*. Toronto: Quality of Life Research Unit, Center for Health Promotion, University of Toronto.

Choi, H. C., & Sirakaya, E. (2005). Measuring residents' attitude toward sustainable tourism: Development of sustainable tourism attitude scale. *Journal of Travel Research, 43*(4), 380–394.

Cummins, R. A. (1992). *Comprehensive quality of life scale* (3rd ed). Melbourne: Psychology Research Centre.

Dillman, D. A. (2000). *Mail and internet surveys: The tailored design method* (2nd ed). New York: Wiley.

Edgerton, R. B. (1990). Quality of life from a longitudinal research perspective. In R. L. Schalock (Ed.), *Quality of life: Perspectives and issues* (pp. 149–160). Washington, DC: American Association on Mental Retardation.

Felce, D., & Perry, J. (1995). Quality of life: Its definition and measurement. *Research in Developmental Disabilities, 16*(1), 51–74.

Ferrans, C. E., & Powers, M. J. (1985). Quality of life index: Development and psychometric properties. *Advances in Nursing Science, 8*(1), 15–24.

Forward, S. (2003). *State of the art report on life quality assessment in the field of transport and mobility*. Linkoeping: Swedish National Road and Transport Institute.

Gursoy, D., & Rutherford, D. G. (2004). Host attitudes toward tourism: An improved structural model. *Annals of Tourism Research*, *31*(3), 495–516.

Heal, L. W., & Sigelman, C. K. (1996). Methodological issues in quality of life measurement. In R. L. Schalock, & G. N. Siperstein (Eds.), *Quality of life: Volume 1-conceptualization and measurement* (pp. 91–104). Washington DC: American Association on Mental Retardation.

Hu, L., & Bentler, P. M. (1999). Cutoff criteria for fit indexes in covariance structure analysis: Conventional criteria versus new alternatives. *Structural Equation Modeling*, *6*(1), 1–55.

Jurowski, C., Uysal, M., & Williams, D. R. (1997). A theoretical analysis of host community resident. *Journal of Travel Research*, *36*(2), 3.

Kline, R. B. (1998). *Principles and practice of structural equation modeling*. New York: The Guilford Press.

Ko, D., & Stewart, W. P. (2002). A structural equation model of residents' attitudes for tourism development. *Tourism Management*, *23*(5), 521–530.

Krueger, R. (1991). *Focus groups: A practical guide for applied research*. Newbury Park, CA: Sage Publications.

Lankford, S. V., & Howard, D. R. (1994). Developing a tourism impact attitude scale. *Annals of Tourism Research*, *21*(1), 121–139.

Liu, J. C., Sheldon, P. J., & Var, T. (1987). Resident perception of the environmental impacts of tourism. *Annals of Tourism Research*, *14*(1), 17–37.

Massam, B. H. (2002). Quality of life: Public planning and private living. *Progress in Planning*, *58*(1), 141–227.

McCool, S. F., & Martin, S. R. (1994). Community attachment and attitudes toward tourism development. *Journal of Travel Research*, *32*(3), 29.

McGehee, N. G., & Anderson, K. L. (2004). Factors predicting rural residents' support of tourism. *Journal of Travel Research*, *43*(2), 131–140.

Perdue, R. R., Long, P. T., & Allen, L. (1987). Rural resident tourism perceptions and attitudes. *Annals of Tourism Research*, *14*(3), 420–429.

Perdue, R. R., Long, P. T., & Allen, L. (1990). Resident support for tourism development. *Annals of Tourism Research*, *17*(4), 586–599.

Rogerson, R. J. (1999). Quality of life, place and the global city. In L. Y. Lim, B. Yuen, & C. Low (Eds.), *Urban quality of life: Critical issues and options*. Singapore: National University of Singapore.

Schalock, R. L. and Siperstein, G. N. (Eds.). (1996). *Quality of life: Volume 1-conceptualization and measurement*. Washington, DC: American Association for Mental Retardation.

Sirgy, M. J., Rahtz, D. R., Cicic, M., & Underwood, R. (2000). A method for assessing residents' satisfaction with community-based services. *Social Indicators Research*, *49*(3), 316–316.

Strauss, A., & Corbin, J. (1990). *Basics of qualitative research*. Newbury Park, CA: Sage Publications.

Taylor, S. J., & Bogdan, R. (1990). Quality of life and the individual's perspective. In R. L. Schalock, & M. J. Begab (Eds.), *Quality of life: Perspectives and issues* (pp. 27–40). Washington, DC: American Association on Mental Retardation.

Teye, V., Sirakaya, E., & Sonmez, S. F. (2002). Residents' attitudes toward tourism development. *Annals of Tourism Research*, *29*(3), 668–688.

Thurstone, L. L. (1947). *Multiple factor analysis*. Chicago: University of Chicago Press.

Wang, Y., & Pfister, R. E. (2008). Residents' attitudes toward tourism and perceive personal benefits in a rural community. *Journal of Travel Research*, *47*(1), 84–93.

Chapter 7
Sustainability Indicators for Managing Community Tourism

HwanSuk Chris Choi and Ercan Sirakaya Turk

Introduction

The purpose of this study is to develop indicators to measure community tourism development within a sustainable framework. In order to develop such objective indicators, this study employed a modified Delphi technique. A panel of 38 academic researchers in tourism provided input into developing the indicators. After three rounds of discussion, the panel members reached consensus on the following set of 125 indicators: political (32), social (28), ecological (25), economic (24), technological (3), and cultural dimensions (13) for community tourism development (Table 7.1). This set of sustainable tourism indicators can serve as a starting point for devising a set of indicators at the local and regional level. Further study shall develop a set of sustainable indicators relying on communities' distinctive characteristics and employing indicator experts from the social and physical sciences and from all stakeholder groups, including residents of the host community, industry experts, government planners, policy makers, and nongovernmental organizations (United Nations, 2001).

In the twentieth century, the globalization of capitalism, the movement of populations, and the advances in transportation and communication technology made tourism into one of the world's largest industries. According to the World Travel and Tourism Council (2004), world tourism receipts will reach approximately $727.9 billion by the end of 2004, with tourism generating more than 214 million jobs and contributing about $5.5 trillion of GDP, 10.4% of the world's total. Because of its ability to create income, taxes, hard currency, and jobs, tourism has become the savior of many communities around the world (Sirakaya, Jamal, & Choi, 2001).

H.C. Choi (✉)
University of Guelph, Guelph, ON, Canada
e-mail: Hwchoi@uoguelph.ca

This is reprinted with permission from Elsevier and the authors from a prior article, Sustainability indicators for managing community tourism, *Tourism Management* 27(2006): 1274–1289. See the link at: http://www.sciencedirect.com/science/journal/tourismmmanagement

M. Budruk, R. Phillips (eds.), *Quality-of-Life Community Indicators for Parks, Recreation and Tourism Management*, Social Indicators Research Series 43, DOI 10.1007/978-90-481-9861-0_7, © Springer Science+Business Media B.V. 2011

Table 7.1 Number of indicators developed from the Delphi study

Dimensions	Number of indicators (issues) developed
Economic dimension	24
Social dimension	28
Cultural dimension	13
Ecological dimension	25
Political dimension	32
Technological dimension	3
Total	125

Response rate – Round 1: 60.6%; Round 2: 62.5%; Round 3: 48.0%.

However, although tourism has brought economic benefits, it has significantly contributed to environmental degradation, negative social and cultural impacts, and habitat fragmentation. Tourism's unplanned growth has damaged natural and socio-cultural environments of many tourism destinations (Domet, 1991; Frueh, 1986; Hall & Lew, 1998; Hidinger, 1996; Mowforth & Munt, 1998; Murphy, 1985; Singh, 1989). These undesirable side-effects have led to the growing concern for conservation and preservation of natural resources, human well-being, and the long-term economic viability of communities (Akis, Peristianis, & Warner, 1996; Butler & Boyd, 2000; Cater, 1993; Hall & McArthur, 1998; Haralambopoulos & Pizam, 1996; Healy, 1994; Mowforth & Munt, 1998; Place, 1995; Richard & Hall, 2000). As decision makers became increasingly aware of the drawbacks of mass tourism, they searched for alternative tourism planning, management, and development options. As a result, the notion of sustainable development (sustainability) emerged as an alternative to the traditional neo-classical model of economic development.

The World Commission on Environment and Development (WCED) issued the first report on sustainability, entitled "Our Common Future," and supported by several international organizations (UNCED, 1992; WCED, 1987; World Conservation Strategy, 1980; WTTC/WTTC/ECC, 1995). The WCED defined sustainable development as development that "meets the needs of the present without compromising the ability of future generations to meet their own needs" (WCED, 1987, p. 43). Following this report, "sustainable development" became an internationally known term and the subject of thousands of books and papers. This concept of sustainable development, although not supported by international agreements and strategies and with much uncertainty as to its underlying theories and processes, became everyone's idea of a universal solution (Redcrift, 1999). It has turned into a "catch-all" term, equally embraced by those whose economic and environmental views are otherwise contradictory (Gowdy, 1999; Hall & Lew, 1998).

It should come as no surprise, then, that sustainable community tourism has had limited practical application in the areas of management, planning, and the monitoring systems at the local level (Butler, 1999). Berry and Ladkin (1997) have argued that the relatively small size of most tourist businesses and the dramatic rise of the sustainability issue have raised serious questions about implementing and monitoring sustainable tourism at local levels. Consequently, individual countries have no clearly defined national policies and strategic reports on sustainable development and its implementation. Neither a common management framework nor indicators exist to systematically track and monitor socio-economic and political changes in communities. According to Weaver and Lawton (1999), indicator studies in tourism are still in their infancy, although the WTO and other organizations are making sporadic efforts to develop them. Without this, universal measurement and monitoring system to guide decision makers in creating tourism policies preventing further degradation and destruction of natural, social, and human resources is inevitable (Sirakaya et al., 2001).

If the changes wrought by tourism upon all aspects of community life are to be effectively tracked, indicators must be based on policy relevance, analytical soundness, and measurability. These indicators can be used in various settings, especially at the local level where it matters the most. The purpose of this study is to develop indicators to measure community tourism development within a sustainable framework using a modified Delphi technique.

Goals of Sustainable Community Tourism

Sustainable development for community tourism should aim to improve the residents' quality of life by optimizing local economic benefits and by protecting natural and built environment and provide a high-quality experience for visitors (Bramwell & Lane, 1993; Hall & Lew, 1998; McIntyre, 1993; Stabler, 1997; UNCED, 1992). Sustainable community tourism guidelines should reflect these goals. Sustainable community tourism should provide a long-term economic linkage between destination communities and industries. It should also minimize the negative effects of tourism on the natural environment, and improve the socio-cultural well-being of the destination communities. Community stakeholders, including governments, tourists, hosts, tour operators, and other tourist-related businesses, must assume the ethical responsibilities and codes of conduct (c.f. Fennell, 1999; Herremans & Welsh, 1999). Decision-making and development processes require multistakeholder involvement at all levels of planning and policy making, bringing together governments, NGOs, residents, industry, and professionals in a partnership that determines the amount and kind of tourism that a community wants. Community managers and planners need to provide educational information and programs (e.g., workshops) to residents, visitors, industry, and other stakeholders in order to raise public and political awareness of the planning and conservation of community tourism resources (Sirakaya et al., 2001). Stakeholders must develop systems that can monitor and adjust planning and destination management.

Dimensions of Sustainable Community Tourism (SCT)

The quest for sustainable tourism indicators must take into account many interpretations of sustainable tourism (Diamantis, 1997; Orams, 1995). Such indicators must be based on the multidimensional nature of sustainable community tourism (Mowforth & Munt, 1998). Traditionally, tourism development contains economic, social or socio-cultural, and environmental dimensions (Mowforth & Munt, 1998). However, recent debates suggest that sustainable community tourism development involves even more dimensions. Sustainable community tourism is comprised of ecological, social, economic, institutional/political, cultural, and technological dimensions at the international, national, regional, and local community levels, and within agriculture, tourism, political sciences, economics, and ecology (Bossell, 1999; Mowforth & Munt, 1998). These dimensions of sustainable community tourism are interdependent and mutually reinforcing (Colby, 1989; Reid, 1995; Slocombe, 1993).

Much of the literature on sustainable tourism has focused on the traditional dimensions (e.g., economic, social, cultural, and ecological dimensions) of tourism and, for this reason these are only briefly discussed in this chapter. Moreover, two additional dimensions, political and technological, are also discussed.

First, there is no doubt that sustainable tourism must be economically feasible because tourism is an economic activity. Economic sustainability, in this regard, implies optimizing the development growth rate at a manageable level with full consideration of the limits of the destination environment. Moreover, the economic benefits from tourism should be fairly well distributed throughout the community. Second, environmental sustainability recognizes that natural resources of the individual community and the world should be no longer viewed as abundant and are, in fact, constantly being depleted. The natural environment must be protected for its own intrinsic value and as a resource for present and future generations. Third, socio-cultural sustainability implies respect for social identity and social capital, for community culture and its assets, and for a strengthening of social cohesiveness and pride that will allow community residents to control their own lives.

According to Pearce (1993), Hall (1994), and McIntosh, Goeldner and Ritchie (1995), sustainable development is a political concept, and therefore achieving the goals of sustainable community tourism depends heavily on the society's political system and power distribution. For example, despite the fact that one goal of sustainable community tourism is improved quality life for local residents in both developed and developing countries, governments control tourism development. As a result, local residents are often excluded from the decision-making process. In order to make sustainable community tourism a reality, residents must have a decision-making role (Gunn, 1994; Hart, 1998; Murphy, 1983; Pigram, 1990; Simmons, 1994).

Although most of the political issues that arise in the course of achieving sustainable community tourism are associated with residents' rights, others include an absence of stakeholder collaboration or community participation, a lack of community leadership, poor regulations, the role of NGOs, and the displacement of resident and external control over the development process by private or foreign investors. As pointed out by Becker, Jahn and Stiess (1999, p. 5), "the main objective *in the political context of sustainability* is to renegotiate the goals of future sustainable *community tourism* and to establish a system of governance that is able to implement policies moving toward sustainability *at all levels*."

Technological advances in transportation, information, and communication systems have heavily influenced the growth of modern tourism (Mathieson & Wall, 1982; Pigram & Wahab, 1997). Marien and Pizam (1997) predicted that technology would play a central role in sustainable community tourism. Initially, the application of a low-impact or environmentally sound technology may minimize the natural, social, and cultural effects of tourism on a destination (Ko, 2001). Next, advanced information technology (e.g., distance education, the World Wide Web, e-mail, and e-commerce) brings benefits to host communities, since these technologies can provide communities with communication networks that permit stakeholders to exchange information (Marien & Pizam, 1997), allow the creation of visitor-education networks and give access to a wider market through the Internet (Milne & Mason, 2000). Moreover, the use of sustainable community tourism (SCT) technologies that are environmentally friendly, socially acceptable, socially appropriate, and managerially supportive have been discussed. These include indigenous technologies (Cater, 1996), soft mobility systems (e.g., car-free environments, bikes, electric and hybrid buses and taxis, high-speed trains) (UNEP, 2001), efficient energy resources (Clayton & Radcliffe, 1996), GIS (Hall, 1996), and eco-labeling (IPS, 2002). Sustainable community tourism demands scientific knowledge and technological support if we are to understand current phenomena, assess/monitor tourism's impact, and provide alternative devices and techniques to avoid future negative outcomes. All dimensions of sustainable community tourism development are interconnected, and serve as basis for the development of SCT indicators.

The Role of Indicators in the Measurement of Sustainable Growth

William Ogburn, in the early twentieth century, was the first to develop statistical measurements to monitor social trends and change. The actual social indicator movement started in the mid-1960s when researchers and politicians became dissatisfied with the quantity and quality of available social information. Since then, use of the term "social indicators" has steadily increased and is now common parlance among managers and researchers who monitor social and biophysical changes (Wallace & Pierce, 1996). As the use of indicators has become widespread, their uses have been expanded to include broad technical indicators (i.e., indirect/direct, descriptive/analytical, and subjective/objective) and discipline-based indicators (e.g., economic indicators, social indicators, tourism indicators, or psychological indicators). Today, many national and international organizations, including the United Nations, International Institute of Sustainable Development (IISD), United Nations Commission of Sustainable Development (UNCSD), Sustainable Seattle, the National Round Table on the Environment and the Economy (NRTEE), United Nations Development Program (UNDP) and the World Bank, Bird Life, New Economics Foundations, and Oxfam International, and the US Interagency Working Group on Sustainable Development Indicators have developed sustainable indicators. These efforts have focused on general development (c.f. physical environment and economy) at the macro level. Not surprisingly, tourism development has not been considered as a main theme of these efforts and most of these indicators cannot be used to monitor tourism development. However, these efforts also had a positive impact on tourism industry and organizations encouraged developing tourism indicators (McCool, Moisey, & Nickerson, 2001; Miller, 2001; Nelson, 1993; Sirakaya et al., 2001; Wallace & Pierce, 1996).

The current literature suggests that considerable progress has been made in planning framework, policy, and tourism-impact research (Bramwell et al., 1998; Garrod & Fyall, 1998; Gunn, 1994; Hunter, 1997; Wahab & Pigram, 1997). However, in spite of the efforts made by several tourism researchers (Miller, 2001; Wallace & Pierce, 1996), several studies have pointed out how little progress has been made in monitoring and measuring sustainable community tourism development (Goodall & Stabler, 1997; Sirakaya et al., 2001; Weaver & Lawton, 1999). This lack of progress may be due to the result of a lack of mutually accepted measurement and monitoring systems. As Moffatt (1996, p. 132) argued, "... at present, no universal agreement on a specific sustainable index has been made ... the lack of a single composite index or useful set of indices poses great difficulties in attempting to monitor whether or not a particular trajectory of a system is on a course of sustainable development or not." In order to assess local tourism development, to guide present action, and to predict future development, decision makers and planners need to know potential monitoring area, data availability, and measuring methods (Manning, 1993).

Monitoring and measurement are the final steps in the planning process, since they can make sustainable tourism development operational. According to Inskeep (1991), assessing the impact of development projects and evaluating their performance during their implementation phases require the use of mitigation measures and indicators. In the past few decades, great efforts have made in developing indicators that will provide better monitoring systems. These efforts have been concerned assessing community development processes in terms of whether these processes successfully met the host communities' needs (Moffatt, 1996). In response to these concerns, social scientists have begun to design an approach that will meet the needs of all stakeholders and form a basis for broad strategic planning. Tourism discussions, forums, conferences, and publications have failed to produce workable indicators within an acceptable framework of sustainable community development. Increasing awareness of the negative impacts of tourism (O'Grady, 1990) and the associated demand for impact-assessment studies, growth-management strategies (Williams & Gill, 1994), and consideration for the carrying capacity of the natural and socio-cultural environment (Getz, 1983; Gunn, 1994) have led to the demand for indicators that can monitor the sustainability of the natural and socio-cultural environment.

In order to apply these complex systems and dimensions of sustainability, sustainability-monitoring systems should be treated differently from traditional mass-tourism approaches (Nieto, 1996). While traditional approaches emphasize the quantitative economic measures of growth and progress, sustainability approaches should make a qualitative improvement to social, natural, and human systems (Gunn, 1994). The welfare of future generations and the balance of wealth among nations have become the center of an ethical debate. If sustainable tourism policies and measures are not established early on to manage the possible negative effects of tourism, initial tourism development can become a political and marketing gimmick that opens the door to unwelcome mass tourism (Bookbinder, Dinerstein, Rijal, Cauley, & Rajouria, 1998; Mowforth & Munt, 1998).

Finally, Manning (1999, p. 180) noted that the objectives of the WTO were to "identify a small set of indicators set which is likely useful in almost any situation; to supplement these with additional indicators known to be useful in particular ecosystem or types of destinations; and to additionally require a scanning process for risks not covered by the aforementioned indicator sets, which produces further indicators critical to the management of the particular site/destination."

Given the complex interrelationships of tourism systems, sustainable indicators for community tourism development (CTD) should be treated differently from traditional indicators and within an adequate development process framework. In order to clarify the goals of sustainable tourism and its indicators, the following guidelines were proposed, based on the works by Liverman, Hanson, Brown, and Meredith (1988), Inskeep (1991), Kuik and Verbruggen (1991), Jamieson (1998), Hart (1998), Bossell (1999), Ross and Wall (1999), Jamal and Getz (1999), Global Reporting Initiative (2000), Sirakaya et al. (2001), United Nations (2001), Veleva and Ellenbecker (2001), and Miller (2001).

- Sustainable tourism strategies must entail ways and means to create adequate policies and proper decision-making processes at all levels of government.
- Sustainable tourism policies should provide workable definitions, principles, implementation strategies, action plans, and a monitoring system of sustainable development for community tourism development with consideration of the entire spectrum of economic, social, cultural, natural, technological, and political environments.
- The context of sustainable tourism is a highly political one involving many stakeholders. Thus, political support in the form of legally binding commitments at the national and regional level is a critical element in obtaining information, funding, education, and expertise.
- In order to reflect the visions and values of a community-based destination, the transparent process of using and evaluating sustainability indicator (SI) development to enable full community participation of all stakeholders must be articulated. These stakeholders should be allowed to influence the direction of current and future CTD.
- SI requires an organizational body (structure and process) to ensure the long-term sustainability of the community-based destination.
- SI should be based on a sustainability framework rather than a traditional development framework, since the latter is either inadequate or inappropriate to measure sustainable growth accurately.

- The number of indicators must be manageable either quantitatively or qualitatively and be easily implementable and in a timely manner at the destination and community level.
- The SI development process requires a systematic approach that has a high degree of reliability, predictive capacity, and integrative ability.
- Clear guidelines on how to select and use SIs in destinations should be proposed and tested.
- SI should be robust, measurable, and affordable, and should provide an integrated view of overall and specific awareness and understanding of past and current performance of community tourism development. It should guide future development and reflect the community's goal.
- SI must serve as an early warning system, not only to prevent the potentially negative impact of tourism development but also to promote sustainable growth.

Methodology

To develop objective indicators, this study employs a Delphi technique. The Delphi method is the best-known qualitative and structured technique for predicting future events by reaching consensus (Poulsen, 1920; Woudenberg, 1991). The panel members who participated in this study were selected in two ways. First, six sustainable tourism experts recommended the potential panelists and another list of potential panelists were drawn from an initial list of 80 authors who had published at least one peer-reviewed paper on sustainable tourism development in journals such as the *Annals of Tourism Research*, *Journal of Travel Research*, *Tourism Management*, and the *Journal of Sustainable Tourism*. Second, the study employed a snowball sampling using 25 internationally recognized tourism scholars identified by the Delphi board. These scholars were asked to provide the names of potential Delphi panel members. The list was then cross-checked for duplication of names. Forty-five potential panel members were identified and 37 participated in the first round of this study. Prior to the second and third rounds, the panel members who agreed to participated in the previous round were contacted and the instrument was sent out to only those panel members willing to share their expertise on developing an indicator set. For this reason, the number of panel members participating in the second and third rounds decreased by 32 and 25, respectively. Thus, compared to some previous studies (Garrod & Fyall, 2000; Miller, 2001; Stein, Clark, & Rickards, 2003), the sample size for this study is relatively small. However, the Delphi study elicits qualitative opinions from panelists who have expertise in the field. According to MacCarthy and Atthirawong (2003), the sampling method of Delphi studies is distinctly different from that of conventional statistical sampling and inferences techniques. The expert panel members were selected based on their expertise and experience in the subject area. Furthermore, Somer, Baker, and Isbell (1984) suggested that limiting the size of panel members makes it easier to control the work generated, since there was a possibility that too much input might bury good data. In this sense, it seemed that the panelists in this study were properly selected with a pool of knowledgeable and representative sustainable tourism expert groups.

Survey Instrument and Data Collection

The survey instrument was based on a thorough review of the relevant literature and discussion with the Delphi board. The first round questionnaire consisted of three sections including questions concerning definitions, principles, and potential indicators. In Section I and II, the panel members were asked to form their own definition and identify the necessary principles of sustainable tourism. Then, based on their definition, panel members were asked to create a list of sustainable tourism indicators useful in monitoring the progress or problem areas in six key dimensions of tourism (i.e., economic, social, cultural, environmental, political/institutional, and technological dimension). In addition, they were asked to answer how the suggested indicator would be operationalized. The panel members were allowed complete freedom to explore the topic and this could elicit the opinion of the panel member on what they considered to be sustainable tourism development indicators.

A group of six academic scholars (three faculty members and three graduate students) reviewed the initial questionnaire. The Delphi study data was entered and analyzed using Excel and SPSS (Statistical Program for Social Sciences, 11.0 version) software. Quantitative analysis included the calculation of mean scores and standard deviation. However, standard deviation is not presented in the tables. The results of the first round – open-ended questions on indicators for sustainable tourism development – were categorized and synthesized for use in the second and third rounds. The Delphi study involved the following rounds.

The first round instrument was sent by regular mail, e-mail, or fax to 37 panel members and 23 (62.2%) were returned. A reminder was sent to panelists who had not returned their questionnaires. The replies were collated into 157 itemized indicators or issues. The second round questionnaire was distributed to 32 panel members and 20 (62.5%) were returned. Respondents were instructed to rate their opinion in terms of agreement, disagreement, or their inability to comment within a 5 point Likert scale (strongly disagree to strongly agree). 3.5 or higher of the items' mean score was used as a cutoff point. In the third round, 25 panel members participated and 12 (48%) were returned. In order to reduce the potential biases produced by *the group thinking* of panel members (Abdel-Fattah, 1997), the statistical feedback (mean score of each item) was provided when the Delphi board distributed this round of questionnaires. To assess the listed indicators, panel members were instructed to assess "soundness" as either "sound" or "not sound." This question was designed to confirm panel members' opinion.

Findings

One hundred twenty-five indicators for six dimensions were identified. The summaries from the second and third rounds of the Delphi study are shown in Tables 7.2, 7.3, 7.4, 7.5, 7.6, and 7.7 along with their mean scores, standard deviation, measurability, and soundness. Table 7.8 presents the top three sustainable indicators of each dimension.

Table 7.2 Indicators for the economic dimension

Key themes	Indicators/issues	Mean	Soundness
Employment	Employment growth in tourism	4.14	0.83
	Unemployment rate	3.80	0.50
	Employment growth in general	3.50	0.50
Income distribution/capital leakage and linkage	Percent of income leakage from the community	4.60	1.00
	Intersectoral linkages/partnership in tourism	4.10	0.92
	Employment and income multipliers on tourism expenditures	3.90	1.00
	Rate change of purchase of local products/value and variety	3.80	1.00
	Import contents	3.78	1.00
	Tourism as percentage of local economy	3.60	0.75
Capital formation in the community/investment	Percent of foreign ownership of tourist establishments	4.40	0.83
	Percent of profit/revenue reinvestment in community development	4.10	0.92
	Availability of local credit to local business	4.05	1.00
	Entrepreneurial opportunities for local residents	3.80	0.92
	External ownership of business in general	3.67	0.92
	Percent of profit/revenue reinvestment in reserved natural and cultural area management and protection	3.56	0.75
Nature of demand	Percent of repeat visitors	4.00	0.67
	Seasonality of tourism/tourist visitation	3.80	0.58
Economic well-being	Comparative ratio of wages in tourism sector to local average wage	3.90	0.92
	New GDP (index of sustainable economic welfare)	3.90	0.75
	Local community economic stability	3.67	0.92
Labor/company and job conditions	Social cost/benefit at community level for examining net benefit to local economy	3.80	0.75
	Existence of an adequate fee structure (e.g., higher entrance fee for tourists and low entrance fee for locals and additional donations)	3.80	0.75
	Equal opportunity employment and promotion to women and local residents	3.70	0.58
Local government income	Hotel/motel tax	3.60	0.50

Mean: Each itemized indicator of each category has been rated by panel members from strongly agree (5) to strongly disagree (1). Cutoff point: 3.5.

Table 7.3 Indicators for the social dimension

Key themes	Indicators/issues	Mean	Soundness
Host community/ residents and stakeholders	Host community satisfaction toward tourism development	4.30	0.83
	Host community attitude toward tourism development	4.20	1.00
	Continuance of traditional activities by local residents	3.70	0.50
	Stress in visitors/host relationship	3.60	0.92
	Resident/nonresident ownership of homes (second homes/part-time residents)	3.60	1.00
	Level of congruency among stakeholders	3.50	0.75
	Resident involvement in tourism industry	4.10	1.00
Social cohesion	Change in social cohesion	3.70	0.67
	Change in community structure evident of a community breakdown and alienation	3.50	0.67
	Change in family cohesion	3.50	0.50
Sex tourism	Sex tourism	3.60	0.75
	Percent employed in sex tourism	3.60	0.42
	Prostitution number and rate in local sex tourism industry	3.60	0.67
	Community attitude toward sex tourism	3.60	0.83
Tourist satisfaction	Tourist satisfaction/attitude toward tourism development	4.00	1.00
Community resource	Degradation/erosion of natural and cultural resource	4.00	0.92
Distribution of resources/power	Shift in social structure (e.g., power shift and its socio-economic implications)	3.80	0.67
	Percent of managerial employment from local residents	3.80	0.83
Community health and safety	Litter/pollution (air, water, etc.)	4.20	1.00
	Overcrowding	4.00	0.92
	Congestion (road)	3.90	0.92
	Crime rate	3.60	0.67
	Public awareness toward value of tourism	3.70	0.75
	Number of incidents of vandalism reported	3.50	0.67
	Community health (availability of health policy related to tourism)	3.50	0.75
	Loss of traditional lifestyle and knowledge via modernization	3.50	0.92
Quality of life in general	Levels of satisfaction with community life in general (QOL)	3.56	0.67

Mean: Each itemized indicator of each category has been rated by panel members from strongly agree (5) to strongly disagree (1). Cutoff point: 3.5.

Table 7.4 Indicators for the cultural dimension

Key themes	Indicators/issues	Mean	Soundness
Building/architecture	Comparability of new construction with local vernacular	4.00	0.75
	Types of building material and décor	3.90	0.75
Cultural (site) management	Cultural sites maintenance level	4.10	0.83
	Availability of cultural site maintenance fund and resource	3.80	0.92
	Commodification	3.80	0.92
	Number of officially designated sites and its management	3.60	0.75
Socio-cultural fabric	Retention of local customs and language	3.90	0.92
	Shift in level of pride in local cultural heritage	3.70	0.92
	Percent satisfied with cultural integrity/sense of security	3.60	0.67
	Loss of authenticity and becoming impersonal	3.50	0.83
Cultural education	Type and amount of training given to tourism employees (guide)	3.80	0.92
	Type of information given to visitors before and during site visits (e.g., tourist in-flight video or public awareness print)	3.60	1.00
	Level of sensitivity of interpretive material and activities pursued	3.50	1.00

Mean: Each itemized indicator of each category has been rated by panel members from strongly agree (5) to strongly disagree (1). Cutoff point: 3.5.

Indicators were selected using the following cutoff point: (a) an indicator score of 3.5 or higher on the agreement rating ("strongly disagree" anchored at 1 to "strongly agree" anchored at 5 in the second round survey because an indicator scoring 3.5 or higher is mid-point between agreement 4 and neutral 3, neither agree nor disagree) and (b) an indicator score of 50% or higher in the "soundness" (yes or no) rating in the third round survey. Three rounds of the survey yielded the 24 indicators (see Table 7.2) for the economic dimension. The researchers agreed that the top priority in indicators for SCT is leakage of tourism-generated income and the strength of locally owned business. Those indicators are "percentage of income leakage" (mean/soundness: $X = 4.60/1.00$) agreement mean/soundness mean, respectively; "percentage of local or foreign ownership" ($X = 4.40/0.83$); "employment and income multipliers" ($X = 3.90/1.00$); "availability of local credit to local business" ($X = 4.05/1.00$); "percentage of income reinvestment" ($X = 4.10/0.92$); and "intersectoral linkage and partnership in tourism" ($X = 4.10/0.92$). On the other hand, tourism researchers rated their levels of agreement as relatively lower in certain areas: "government income and general trend including hotel model tax" ($X = 3.60/0.50$),

Table 7.5 Indicators for the ecological dimension

Key themes	Indicators/issues	Mean	Soundness
Loss of renewable resources	Air quality index	4.44	1.00
	Amount of erosion on the natural sites	4.22	1.00
	Number of good air quality days	4.13	1.00
	Frequency of environmental accidents related to tourism	4.13	0.92
	Number of contaminated sites	4.11	0.75
Rate of ecosystem destruc-tion/degradation	Level of protection (IUCN classification, e.g., parks beaches, species, fragile ecosystem)	4.22	0.83
	Per capita water/energy consumption data	4.00	1.00
	Percent of are under protection designation or environmentally managed	3.89	1.00
Assessment of environmental impacts of tourism activity	Continues use of environmental impact assessment	4.22	0.83
	Natural environment accounting and life cycle analysis	4.11	0.83
	Number of endangered species	3.67	1.00
Reuse/recycling rates	Availability, size, and condition of urban forest	4.00	0.92
	Timber growth/removal	4.00	0.92
	Renewable resources used	4.00	0.92
	Recycling rate	3.89	0.75
	Fisheries utilization	3.89	0.83
Health of human population (residents/visitors)	Formal control required over development of sites and use densities	4.11	0.75
	Type and amount of environmental education training given to employee (guide)	4.00	0.92
	Use intensity	4.00	0.92
	Per capita discharge of waste water	3.89	1.00
	Group size in sites (carrying capacity)	3.89	0.83
	Per capita discharge of solid waste	3.75	1.00
Loss of nonrenewable resources	Level of losing vegetation	4.11	1.00
	Stress level and loss of endangered species	4.00	1.00
	Site attractivity	3.67	0.83

Mean: Each itemized indicator of each category has been rated by panel members from strongly agree (5) to strongly disagree (1). Cutoff point: 3.5.

"unemployment rate" ($X = 3.80/0.50$), and "employment growth in general" ($X = 3.50/0.50$).

In the social dimension, 27 indicators/issues (see Table 7.3) were identified. The panel members heavily favored "satisfaction and attitude of hosts and guests" and "community health and safety." These include "host-community satisfaction"

Table 7.6 Indicators for the political dimension

Key themes	Indicators/issues	Mean	Soundness
Local oriented control policy	Availability of development control policy	4.44	1.00
	Legal compliance (prosecutions, fines, etc.)	4.33	1.00
	Presence of tourism authority or planner in local community	4.13	0.92
	Strength and duration of local advisory and planning group	4.11	0.75
	Percent of foreign/external ownership of tourism establishment	4.11	1.00
	Incorporation and implementation of local idea in community/site management	3.89	0.83
	Intersectoral linkages at local/regional/national level	3.89	0.75
	Local environmental NGOs	3.78	0.92
Political participation	Local resident participation in planning process	4.44	1.00
	Stakeholder collaboration	4.38	0.75
	Level of cooperation among stakeholder groups	4.33	0.92
	Public–private sector partnership	4.00	0.75
	Availability of resident advisory board	3.89	0.92
	Awareness of meaning and implications of sustainable (define) tourism	3.78	0.75
Local planning policy	Availability of air and water pollution and waste management and policy	4.44	1.00
	Availability and level of land zoning policy	4.44	1.00
	Tourism-related master plan	4.33	0.92
	Inclusion of tourism into a community planning process as one of major components	4.11	0.75
	Formal evaluation of implementation and/or process of sustainable tourism plan	4.11	1.00
	Availability of visitor safety, security, and health policy and funding	4.00	0.92
	Existence of sustainable tourism plan	4.00	0.92
	Building permits issued (overall and directly tourism related)	3.89	0.83
	Code of ethics	3.67	0.92
Political supports at all levels of government	Incorporation and implementation of local ideas in community/site management	4.22	0.92
	Attitude of local political and NGO leaders toward development and conservation	4.22	0.92
	Two-way communication between residents and local government	4.13	0.92
	Level of support for conservation/development projects at local level	4.11	0.58

Table 7.6 (continued)

Key themes	Indicators/issues	Mean	Soundness
	Availability and types of performance-based incentive programs	4.00	0.92
	Level of support for conservation/development projects at regional level	4.00	0.75
	Availability, type, and level of committee/training program	3.88	0.75
	Level of support for conservation/development projects at national level	3.75	0.75
	Availability of affordable funding resources	3.75	0.67

Mean: Each itemized indicator of each category has been rated by panel members from strongly agree (5) to strongly disagree (1). Cutoff point: 3.5.

($X = 4.30/0.83$); "attitudes toward tourism development" ($X = 4.20/1.00$); "residents' involvement in tourism industry" ($X = 4.10/1.00$); "change in tourist satisfaction/attitude toward tourism" ($X = 4.00/1.00$); "littering" ($X = 4.20/1.00$); "degradation of natural and cultural resources" ($X = 4.00/0.92$); and "crowding" ($X = 4.00/0.92$). Panel members reached agreement on 13 cultural indicators (see Table 7.4) and rated cultural (site) management as the most important area. Some of the other top cultural indicators were "comparability of new construction with local vernacular" ($X = 4.00/0.75$); "cultural site maintenance level" ($X = 4.10/0.83$); and "retention of local customs and language" ($X = 3.90/0.92$).

The "environmental" category (see Table 7.5) produced 26 indicators. "Loss of renewable resources" consists of five indicators, including "air quality index" ($X = 4.44/1.00$), "amount of erosion on the natural sites" ($X = 4.22/1.00$), and "number of contaminated sites" (4.11/0.75). Other indicators favorably supported by panel members include "level of protection" ($X = 4.22/0.83$); "continue use of environment impact assessment" ($X = 4.22/0.83$); "availability, size, and condition of urban forest" ($X = 4.00/0.92$); "formal control required over development of sites and use densities" ($X = 4.11/0.75$); and "level of losing vegetation" ($X = 4.11/1.00$).

Table 7.7 Indicators for the technological dimension

Indicators/issues	Mean	Soundness
Accurate data collection and tourism information change	3.89	0.83
Adoption and use of new and low-impact technologies	3.88	0.83
Benchmarking – generic and competitive (input/output efficiency)	3.50	1.00

Mean: Each itemized indicator of each category has been rated by panel members from strongly agree (5) to strongly disagree (1). Cutoff point: 3.5.

Table 7.8 Top three objective indicators of each dimension

Ranking	Economic dimension	Ranking mean score
1	Availability of local credit to local business	4.7
2	Employment growth in tourism	6.3
3	Percent of income leakage out of the community	7.0
	Social dimension	
1	Resident involvement in tourism industry	2.7
2	Visitor satisfaction/attitude toward TD	3.0
3	Litter/pollution	4.0
	Cultural dimension	
1	Availability of cultural site maintenance fund and resources	3.3
2	Type and amount of training given to tourism employees (guide)	3.7
3	Types of building material and décor	4.7
	Ecological dimension	
1	Air quality index	1.0
2	Amount of erosion on the natural site	2.3
3	Frequency of environmental accidents related to tourism	3.0
	Political dimension	
1	Availability and level of land zoning policy	1.3
2	Availability of air and water pollution and waste management and policy	1.7
3	Availability of development control policy	2.3
	Technological dimension	
1	Accurate data collection	1.3
2	Use of low impact technology	1.7
3	Benchmarking	1.7

Ranking mean scores of all indicators developed ranged from 1 to 32.5.

Tourism researchers recognized that tourism development is a political issue. Although past studies have not identified politics as a key issue in tourism development, the political dimension, with four important areas and 32 indicators (see Table 7.6), were identified including "control policy" ($X = 4.44/1.00$), "resident participation" ($X = 4.44/1.00$), "land zoning policy" ($X = 4.44/1.00$), and "effective communication between residents and local government" ($X = 4.13/0.92$). Finally, only three technological indicators were identified. These were "accurate data collection and tourism information exchange" ($X = 3.89/0.83$), "adoption and use of new and low-impact technologies" ($X = 3.88/0.83$), and "benchmarking" ($X = 3.50/1.00$).

Discussion and Conclusions

After three rounds, the indicators that panel members provided were classified into three categories: checklist indicators, core indicators, and indicator issues. Checklist indicators can help local communities examine the current condition of their resources. However, the checklist indicator does not measure developmental impact or progress, but instead addresses the availability of necessary resources that meet the general requirements for effective planning of sustainable community tourism. The Delphi panel members provided 26 itemized checklist indicators (see Appendix) in the areas of human resources; policy and regulations; funding; and planning and education. These four components are key conditions for sustainable tourism development. However, the local government does not often have enough resources to meet all checklist indicators. Since the support of national and international government is crucial, these indicators represented the combined efforts of all levels of government.

First, Moseley (2002) and Pearce, Moscardo, and Ross (1996) have argued that the community self-determination and active community involvement are essential elements of sustainable tourism. In other words, sustainable tourism development should be planned and managed by community stakeholders. In particular, local governments should solicit their residents' broad and direct participation, which can influence decision making and guarantee to all stakeholders a fair distribution of benefits. In order to maximize community participation, residents must have sufficient access to various communication channels such as Internet, e-mail, and mass and print media. This study result also shows that local communities should form advisory boards made up of representatives from all stakeholder groups, including residents: an effective way for the public to participate in the planning and decision making. Additionally, proper staff, competent authority, and proper technical support among all stakeholders are required for the smooth flow of information.

Second, the study findings indicate that sustainable tourism at the local community level requires at least a development control policy, a natural environment-related policy, and a security and visitor safety policy. Legal compliance regulations (enforced by, for example, fines), land zoning regulations, and development regulations are also required. Often, local governments, especially in developing countries, do not have development-related policies and regulations, and thus tourism should be developed and operated within the regional and national context. Furthermore, in STD policy and regulation formulations, a broadly emerging concern related to responsibility and ethics should be carefully examined and the value of community and environment (both natural and man-made) must be reflected (McIntyre, 1993). Finally, legislative reform should encourage more responsible corporate behavior.

Third, the study panels noted that tourism requires further investment to construct and improve infrastructure, to maintain natural and cultural resources, and to support the local industry. The lack of funding is a chronic problem in tourism development, particularly in rural community destinations and in developing countries.

The last but not least dimension of checklist indicators is planning and education. Tourism planning is a necessary condition to achieve the goals of community tourism development (Gunn, 1994). Furthermore, as Murphy emphasized, residents are willing to participate in decision-making process for their community current and future development. Accordingly, communities can help proper planning by ensuring that all participants in the planning process are well informed about the pertinent issues: providing residents with complete and relevant information and fostering directed learning are two ways of achieving this.

Education and training programs for visitors and other stakeholders are also a crucial tool for delivering accurate interpretations and information about a region/destination. The primary objectives of education are to increase visitors' and other stakeholders' awareness of the fragile nature of local community environment, both man-made and natural (i.e., code of conduct/ethics), and change their behavior and attitude. Therefore, before visiting a region, tourists should understand the culture, society, and nature of the destination through self-educated learning materials and programs such as the code of ethics.

Indicators are often defined as "a quantitative account of a complex situation or process" (OECD, 1997, p. 14) and provide information in a simplified, numerical, and communicative form (OECD, 1997; Peterson, 1997). In this study, several panel members questioned how to measure broadly identified indicators. Fortunately, most of these indicators can be quantified, simplified, and transformed into communicable information or data. For example, "change in social cohesion" can be measured by developing a scale that evaluates perceived cohesion in community psychology.

Similar concepts such as "sense of community" and "place/community attachment" can be substituted to measure change in social cohesion. Furthermore, social cohesion indicators can be adapted from social indicators. According to Rossi and Gilmartin (1980, p. 7), subjective indicators reflect the "comments people make about their emotions, attitudes, attributes and personal evaluation." On the other hand, panel members identified indicators, such as "commodification," as difficult to measure, though it might be addressed by indicators such as "number of conservation plans implemented," "number of sites under threat due to fast commodification," and "number/type of local events/festivals commodified." These examples can explain only the tip of the iceberg of commodified culture, history, and resident identity (Greenwood, 1977). These indicators should be treated as qualitative (subjective) indicators since objective measurement of commodification indicators may fail to explain current progress and changes. Indicator information can be collected from survey questionnaires or in-depth interviews, structured interviews, focus groups, and case studies. The measurement of these indicators can take the form of statements or statistical data which can be used to assess the progress of community tourism development toward sustainable development. Meadows (1998) supported the applicability of the subjective indicator by insisting that "the scientific worldview is just one way to see the world" (p. 9).

Sustainability indicators for community tourism development differ from traditional development indicators because of the interrelationships of natural and cultural resources and stakeholders. Conventional economic indicators such as gross domestic product (GDP) fail to capture the crucial aspects of sustainable economic growth. In other words, GDP, because it takes cultural and natural resources for granted, does not consider the pace and extent to which natural resources are consumed and cultural resources ruined to produce that GDP. As the United Nations has suggested, environmental degradation and depletion should be subtracted from GDP as a cost of production (see United Nations, 2002, Chapter IV, Section D).

Panel members agreed that conventional indicators like GDP could not be objective measures of sustainable economic growth in the development of community tourism (Daly & Cobb, 1989). In tourism, cultural, natural, and other environmentally related resources have been considered in a similar manner; however, the sustainability framework for community tourism development requires full-cost accounting. Another emerging indicator that warrants study in less-developed and less-Westernized countries and communities is the use of indigenous knowledge (IK), also referred to as traditional or local knowledge (Mascarenhas & Veit, 1995).

According to the International Institute of Rural Reconstruction (1996), IK is "the knowledge that people in a given community have developed over time, and continue to develop. It is based on experience, often tested over centuries of use, adapted to local culture and environment, and is dynamic" (http://www.panasia.org.sgl). For example, Reichel (1993) describes shamanistic concepts of environmental accounting in *Columbian Amazonian* in which religion and ritual are used as key indicators to protect the natural environment. For instance, based on traditional analysis of ecosystem boundaries where they live (i.e., land, water, and air) they monitor their impacts on nature. Then, a systematic route for movement (e.g., hunting and fishing routes) and ecosystem boundaries is reestablished. If necessary, they limit to access to certain regions, ban certain activities, and disallow to utilize certain resources. Within this shamanistic concept of sustainable development, community as a whole has the responsibility to preserve and protect natural environment as well as to utilize resources wisely. In this context, rituals are roled as exchanging information on consumption of natural environment. Furthermore, IK will help planners and policy makers with Western perspectives understand local systems and communication processes.

Conventional tourism development strategies with their potential for damage pose a threat to host communities. In order to enhance the local residents' quality of life and to achieve an optimal level of development for rural communities, an effective and efficient monitoring system of tourism impact is *sine qua non*. Managing development and decision-making processes requires new ways of assessing progress, namely, the development of indicators for sustainable community tourism. Hitherto, little research has been done to develop such indicators as monitoring tools (McCool et al., 2001).

Although the debate on sustainable tourism continues among academics and within the industry, sustainable tourism practices have been adopted worldwide. Additionally, scholars who criticize sustainable tourism could not provide alternative recommendations for the tourism industry. Thus, this study expands the current understanding of the field of sustainable community tourism by developing the monitoring system, sustainability indicators for managing development of community tourism.

Some empirical studies create indices of sustainable community development at the local level. Because of their failure to incorporate all dimensions of sustainable development, these studies are incomplete (see McCool et al., 2001; Miller, 2001). The present study bridges this gap. While most available monitoring indices focus on dominant economic, physical, and ecological dimensions, this study extends the spectrum by including the social, cultural, technological, and political dimensions.

In short, a holistic approach to sustainable tourism development should be ecologically responsible, socially compatible, culturally appropriate, politically equitable, technologically supportive, and, finally, economically viable for the host community. To realize this concept at the local level, an effective set of indicators for rural communities is indispensable. This study presents a set of sustainable indicators that serve as a planning tool in the sustainable tourism developmental process for rural communities. Monitoring the impact of tourism at the local level can be more useful because communities are where tourism takes place. The 125 indicators can be used to monitor the impact of tourism on rural communities.

Sustainability, if viewed as a social, cultural, economic, ecological, technological, and political phenomenon, must be studied within a comprehensive planning framework supported by interdisciplinary research, management, monitoring, and feedback. If public and private sector leaders agree to see sustainability as constituting the ultimate societal goal, then the focus of the debate can shift to the crafting of sustainability indicators. Thus, sustainability criteria could be tailored to each region. For example, the sustainability goal of a small region with a large population would differ significantly from that of a large community with a small population (c.f. distinctive characteristics, availability of funding, level of support by the regional and national government, and availability of existing data). The leaders of both communities have sustainability in common, but their differing circumstances would elicit different approaches. Furthermore, not all indicators need to be considered in developing an indicator set for a community. Each community should adopt only the indicators it needs to monitor tourism development.

This study solicited the opinions of specialists in sustainable tourism. Further study should operationalize the indicators developed and then create a set of sustainable indicators. It needs to employ experts from economics, the social sciences, and the physical sciences and all stakeholder groups, including residents of the host community, industry experts, government planners, policy makers, and nongovernmental organizations (United Nations, 2001). In order to be more applicable, micro (community, local, or regional) indicators should be developed with the support of regional, national, and/or international governmental organizations. In further development of sustainable community indicators, involving residents is crucial because they are a major stakeholder group. Resident involvement is the philosophical basis of sustainable community tourism. Furthermore, educating stakeholder groups should be a top priority because one of the major failures in implementing indicators at the local level has been a lack of awareness and participation among stakeholders (UK Department of Environment, Transport, & Regions, 2000). The review of the literature shows that only a few sustainable indicators for SCT were tested in a destination setting. In order to build the efficiency and effectiveness of indicators that monitor the impact of tourism on natural and cultural resources and host communities, these indicators should be tested in a real rural community setting.

The implementation of indicators assists destination managers and planners to achieve sustainable community tourism goal, and to alert them to possible social trends, to changes in the host community, and to the potential negative impact of tourism on natural and cultural resources. Furthermore, effective indicators provide "decision makers of host communities with information that enables them to identify, evaluate and make timely decisions on critical changes being caused by tourism to the natural environment, communities and other resources in the destination" (Sirakaya et al., 2001, p. 425).

In tourism and in other fields, objective indicators are a central tenet in monitoring studies. Interestingly, this study found that some indicators suggested by academic experts fall into subjective indicator category. While the objective indicator can be defined as a tool measuring income, employment rate, and number of visitors, the subjective indicators evaluate attitudes, perception, and satisfaction of community residents. Subjective indicators are usually measured using survey research or qualitative research methods. Studies have indicated that the residents' role is crucial. Consequently, it is important to understand and continually assess their opinions, attitudes, and perceptions of tourism development within a sustainable framework (Johnson, Snepenger, & Akis, 1994; Sheldon & Abenoja, 2001). In this regard, further study should develop subjective indicators to measure resident attitudes, satisfaction, and perception of tourism development and should test developed indicators to extend the current body of tourism impact literature. The evaluation of implementation using both objective and subjective indicators will help create strong monitoring systems.

Acknowledgments   We wish to express our sincere gratitude to the study panel members, but not limited to, Dr. Stephen L.J. Smith, Dr. Muzafar Uysal, Dr. Steve McCool, Dr. Ted Mannning, Dr. Grodon Dickinson, Dr. David Weaver, Dr. Ginger Smith, Dr. Colin Hunter, Dr. Bob Mckercher, Dr. Brian Goodall, Dr. David Timothy Duval, Dr. Erlet Cater, and Dr. Cooper. Our thanks are extended to the expert reviewers, especially Dr. Turgut Var, Dr. Tazim Jamal, Dr. Scott Shaffer, and Professor Hesus Hinojosa.

Appendix

Checklist Indicators

Area	Indicators
Human resource	1. Resident advisory board
	2. Tourism authority or planner in your community
	3. Permanent staffs in information or visitor center
	4. Community-controlled reinvestment committee
Policy and regulations	1. Development control policy
	2. Legal related compliance regulations
	3. Air and water pollution and waste management and policy
	4. Land zoning policy
	5. Visitor safety, security, and health policy
	6. Formal control required over development of sites and use densities
	7. Local purchasing policy
	8. Mandated use of environmental impact assessment
	9. Local/regional/national protected area classification
Funding	1. Local reinvestment fund
	2. Local credit to local entrepreneurship
	3. Cultural and natural site maintenance fund and resource
	4. Visitor safety, security, and health policy related funding
	5. Existence of an adequate fee structure
Planning and education	1. Tourism development master plan
	2. Existence of sustainable tourism development plan
	3. Tourism as one of major components in a community planning process
	4. Continuous monitoring programs delivered
	5. Performance-based incentive programs
	6. Training programs for local stakeholder groups
	7. Use of new and/or low-impact technologies
	8. Code of ethics at all levels of government

References

Abdel-Fattah, N. M. (1997). *Road freight privatization in Egypt: A comparative with Great Britain & Hungary*. Unpublished Ph. D. Dissertation, University of Plymouth.

Akis, S., Peristianis, N., & Warner, J. (1996). Residents' attitudes to tourism development: The case of Cyprus. *Tourism Management, 17*(7), 481–494.

Becker, E., Jahn, T., & Stiess, I. (1999). Exploring uncommon ground: Sustainability & social sciences. In E. Becker & T. Jahn (Eds.), *Sustainability & the social sciences* (p. 5). Hamburger: UNESCO & Institute for Social-Ecological Research (ISOE).

Berry, S., & Ladkin, A. (1997). Sustainable tourism: A regional perspective. *Tourism Management, 18*(7), 433–440.

Bookbinder, M. P., Dinerstein, E., Rijal, A., Cauley, H., & Rajouria, A. (1998). Ecotourism's support of biodiversity conservation. *Conservation Biology, 12*(6), 1399–1404.

Bossell, H. (1999). *Indicator for sustainable development: Theory, method, & application*. Manitoba: International Institute for Sustainable Development (IISD).

Bramwell, B., Henry, I., Jackson, G., Prat, A., Richards, G., & Van der Straaten, J. (1998). *Sustainable tourism management: Principles & practices* (2nd ed.). Tilburg: Tilburg University Press.

Bramwell, B., & Lane, B. (1993). Sustainable tourism: An evolving global approach. *Journal of Sustainable Tourism, 1*(1), 1–5.

Butler, R. W. (1999). Sustainable tourism – A state of the art review. *Tourism Geographies, 1*, 7–25.

Butler, R. W., & Boyd, S. (2000). *Tourism & national parks*. Chichester: Wiley.

Cater, E. A. (1993). Ecotourism in the third world: Problems for sustainable tourism development. *Tourism Management, 14*(2), 85–89.

Cater, E. A. (1996). Ecotourism in the Caribbean: A sustainable option for Belize & Dominica? In L. Briguglio, R. W. Butler, D. Harrison & W. L. Filho (Eds.), *Sustainable tourism in islands & small states: Case studies* (pp. 122–146). New York: Pinter Publisher.

Clayton, A. M. H., & Radcliffe, N. J. (1996). *Sustainability*. London: Earthscan Publication Ltd.

Colby, M. (1989). *The evolution of paradigms of environmental management in development*. World Bank Strategic Planning & Review Discussion Papers 1, Washington, DC: The World Bank.

Daly, H. E., & Cobb, J. (1989). *For the common good: Redirecting economy toward community, the environment, & a sustainable future*. Boston: Beacon Press.

Diamantis, D. (1997). The development of ecotourism & the necessity of using environmental auditing in its planning agenda. *General Technical Report* (pp. 19–23), Northeastern Forest Experiment Station, USDA Forest Service. Northeastern Forest Experiment Station, NE-232, Radnor: USDA Forest Service No.

Domet, R. (1991, March). The Alps are dying. *World Press Review, 38*, 54–55.

Fennell, D. A. (1999). *Ecotourism: An introduction*. New York: Routledge.

Frueh, S. (1986). *Problems in a tropical paradise: The impact of international tourism on Cancun, Mexico*, Unpublished Master Thesis, University of South Carolina.

Garrod, B., & Fyall, A. (1998). Beyond the rhetoric of sustainable tourism? *Tourism Management, 19*(3), 199–212.

Garrod, B., & Fyall, A. (2000). Managing heritage tourism. *Annals of Tourism Research, 27*(3), 682–708.

Getz, D. (1983). Capacity to absorb tourism concepts & implications for strategic planning. *Annals of Tourism Research, 10*, 239–263.

Global Reporting Initiative. (2000). Sustainability reporting guidelines on economic, environment al and social performance. Accessed September 5, 2002, from http://www.globalreporting.org

Goodall, B., & Stabler, M. J. (1997). Principles influencing the determination of environmental standards for sustainable tourism. In M. J. Stabler (Ed.), *Tourism & sustainability. Principles to practice* (pp. 279–304). Oxford: CAB International.

Gowdy, J. (1999). Economic concepts of sustainability: Relocating economic activity within society & environment. In E. Becker & T. Jahn (Eds.), *Sustainability & the social sciences* (pp. 162–181). Hamburger: UNESCO & ISOE.

Greenwood, D. J. (1977). Culture by the pound: An anthropological perspective tourism as cultural commoditization. In V. L. Smith (Ed.), *Hosts & guests: The anthropology of tourism* (pp. 129–137). Philadelphia: University of Pennsylvania Press.

Gunn, C. A. (1994). *Tourism planning: Basics, concepts, cases* (3rd ed.). Washington, DC: Taylor & Francis.

Hall, C. M. (1994). *Tourism & politics: Policy, power, & place*. Chichester: Wiley.

Hall, P. A. V. (1996). Use of GIS based for sustainable development: Experience & potential. *A Newsletter of Institute of Public Finance and Infrastructure finance (IFIP) working group, 6*(2). Accessed April 29, 2001, from http://www.qub.ac.uk/mgt/papers

Hall, C. M., & Lew, A. A. (1998). The geography of sustainable tourism development: Introduction. In C. M. Hall & A. A. Lew (Eds.), *Sustainable tourism: Geographical perspectives* (pp. 1–24). New York: Addison Wesley Longman Ltd.

Hall, C. M., & McArthur, S. (1998). *Integrated heritage management: Principles & practice*. London: The Stationery Office.

Haralambopoulos, N., & Pizam, A. (1996). Perceived impacts of tourism. The case of Samos. *Annals of Tourism Research, 23*(3), 503–526.

Hart, M. (1998). *Indicators of sustainability*. Accessed July 14, 2000, from http://www.subjectmat ters.com/indicators

Healy, R. G. (1994). Tourist merchandise' as a means of generating local benefits from ecotourism. *Journal of Sustainable Tourism, 2*(3), 137–151.

Herremans, I. M., & Welsh, C. (1999). Developing & implementing the company's eco-tourism mission statement. *Journal of Sustainable Tourism, 7*(1), 48–76.

Hidinger, L. A. (1996). Measuring the impact of ecotourism on animal populations: A case study of Tikal National Park, Guatemala. *Yale Forestry & Environment Bulletin, 99*, 49–59.

Hunter, C. (1997). Sustainable tourism as an adaptive paradigm. *Annals of Tourism Research, 24*(4), 850–867.

Inskeep, E. L. (1991). *Tourism planning: An integrated & sustainable development approach*. New York: Van Nostrand Reinhold.

Institute for Policy Studies (IPS). (2002). *Certification & eco-labeling: Amid confusion, consensus is emerging around standards for green tourism*, IPS. Accessed July 20, 2002, from http://www.ips-dc.org/ecotourism/pppr.htm

International Institute of Rural Construction (IIRR). (1996). *Recording & using indigenous knowledge: A manual*. Silang: IIRR.

Jamal, T., & Getz, D. (1999). Community roundtables for tourism-related conflicts: The dialectics of consensus and process structures. *Journal of Sustainable Tourism, 7*(3), 356–378.

Jamieson, D. (1998). Sustainability & beyond. *Ecological Economics, 24*, 183–192.

Johnson, J. D., Snepenger, D. J., & Akis, S. (1994). Host resident perceptions of tourism in a transitional rural economy. *Annals of Tourism Research, 21*(3), 629–642.

Ko, T. G. (2001). *Assessing progress toward sustainable tourism development*. Unpublished Thesis, Sydney, Australia: University of Technology.

Kuik, O., & Verbruggen, H. (Eds.) (1991). *In search of indicators of sustainable development*. Dordrecht: Kluwer Academic Publishers.

Liverman, D. M., Hanson, M. E., Brown, J. B., & Meredith, R. W., Jr. (1988). Global sustainability: Toward measurement. *Environmental Management, 12*(2), 133–143.

MacCarthy, B. L., & Atthirawong, W. (2003). Factors affecting location decisions in international operations – a Delphi study. *International Journal of Operations & Production Management, 23*(7), 794–818.

Manning, E. W. (1993). *What tourism managers need to know: A practical guide to the development and use of indicators of sustainable tourism*. Madrid: World Tourism Organization.

Manning, T. (1999). Indicators of tourism sustainability. *Tourism Management, 20*(1), 179–181.

Marien, C., & Pizam, A. (1997). Implementing sustainable tourism development through citizen participation in the planning process. In S. Wahab & J. Pigram (Eds.), *Tourism, development and growth* (pp. 164–178). London: Routledge.

Mascarenhas, O., & Veit, P. G. (1995). *Indigenous knowledge in resource management: Irrigation in Msanzi, Tanzania*. Baltimore: World Resource Institute.

Mathieson, A., & Wall, G. (1982). *Tourism: Economic, physical & social impacts*. Harlow: Longman.

McCool, S. F., Moisey, R. N., & Nickerson, N. P. (2001). What should tourism sustain? The disconnect with industry perceptions of useful indicators. *Journal of Travel Research, 40*(4), 124–131.

McIntosh, R. W., Goeldner, C. R., & Ritchie, J. R. B. (1995). *Tourism: Principles, practices, and philosophies* (7th ed.). New York: Wiley.

McIntyre, G. (1993). *Sustainable tourism development: Guide for local planners*. Madrid: World Tourism Organization.

Meadows, D. (1998). *Indicators & information systems for sustainable development*. A Report to the Balaton Group, September, 1998.

Miller, G. (2001). The development of indicators for sustainable tourism: Results of a Delphi survey of tourism researchers. *Tourism Management, 22*, 351–362.

Milne, S., & Mason, D. (2000). *Tourism, I.T. & community development.* 4th New Zealand Tourism & Hospitality Conference: New Zealand Tourism: Meeting Challenges & Seizing Opportunities, Auckland, New Zealand: New Zealand Tourism and Hospitality Conference.

Moffatt, I. (1996). *Sustainable development principles, analysis & policies.* New York: Parthenon Publishing.

Moseley, M. J. (2002). *Sustainable rural development: The role of community involvement & local partnership.* NATO Advanced Research Workshop. Krakow, Poland, November 2002.

Mowforth, A., & Munt, I. (1998). *Tourism & sustainability: New tourism in the third world.* London: Routledge.

Murphy, P. E. (1983). Perceptions & attitudes of decision-making groups in tourism centers. *Journal of Travel Research, 21*, 8–12.

Murphy, P. E. (1985). *Tourism: A community approach.* New York: Methuen.

Nelson, J. G. (1993). Are tourism growth and sustainability objectives compatible? Civil, assessment, informed choice. In J. G. Nelson, R. W. Butler, & G. Wall (Eds.), *Tourism and sustainable development: Monitoring, planning, and managing* (pp. 259–268). Waterloo, ON: Department of Geography, University of Waterloo.

Nieto, C. C. (1996). Toward a holistic approach to the ideal of sustainability. *Society for Philosophy & Technology, 2*(2), 41–48.

Orams, M. B. (1995). Towards a more desirable form of ecotourism. *Tourism Management, 16*(1), 3–8.

Organization for Economic Co-operation and Development (OECD). (1997). *Better understanding our cities: The role of urban indicators.* Paris: OECD.

O' Grady, R. (1990). Acceptable tourism. *Contours (Bangkok), 4*(8), 9–11.

Pearce, P. L. (1993). Tourist-resident impacts: Examples, explanations & emerging solutions. In W. F. Theobald (Ed.), *Global tourism: The next decade* (pp. 103–113). Oxford: Butterworth-Heinemann.

Pearce, P. L., Moscardo, G. M., & Ross, G. F. (1996). *Understanding & Managing the Tourist-Community Relationship.* London: Elsevier.

Peterson, T. (1997). *Sharing the earth: The rhetoric of sustainable development.* Columbia, SC: University of South Carolina Press.

Pigram, J. J. (1990). Sustainable tourism – Policy considerations. *Tourism Studies, 1*(2), 2–9.

Pigram, J. J., & Wahab, S. (1997). The challenges of sustainable tourism growth. In S. Wahab & J. J. Pigram (Eds.), *Tourism development & growth: The challenge of sustainability* (pp. 3–16). New York: Routledge.

Place, S. E. (1995). Ecotourism for sustainable development: Oxymoron or plausible strategy? *GeoJournal, 35*(2), 161–174.

Poulsen, F. (1920). *Delphi.* London: Gyldendal.

Redcrift, M. (1999). Sustainability & sociology: Northern preoccupations. In E. Becker & T. Jahn (Eds.), *Sustainability & the social sciences* (pp. 59–73). Hamburger: UNESCO & ISOE.

Reichel, E. (1993). Shamanistic modes for environmental accounting in the Colombian Amazon: Lessons from indigenous ethno-ecology for sustainable development. *Indigenous Knowledge & Development Monitor, 1*(2). Accessed March 1, 2003, from http://www.nuffic.nl

Reid, D. (1995). *Sustainable development: An introductory guide.* London: Earthscan Publications.

Richard, G., & Hall, D. (2000). *Tourism & sustainable community development.* New York: Routledge.

Ross, S., & Wall, G. (1999). Ecotourism: Towards congruence between theory and practice. *Tourism Management, 20*(1), 123–132.

Rossi, R. J., & Gilmartin, K. J. (1980). *The handbook of social indicators: Source, characteristics, & analysis.* New York: Garland STPM Press.

Sheldon, P. J., & Abenoja, T. (2001). Resident attitudes in a mature destination: The case of Waikiki. *Tourism Management, 22*(5), 435–443.

Simmons, D. G. (1994). Community participation in tourism planning. *Tourism Management, 15*(2), 98–108.

Singh, S. C. (1989). *Impact of tourism on mountain environment*. India: Research India Publications.

Sirakaya, E., Jamal, T., & Choi, H. S. (2001). Developing tourism indicators for destination sustainability. In D. B. Weaver (Ed.), *The encyclopedia of ecotourism* (pp. 411–432). New York: CAB International.

Slocombe, D. S. (1993). Environmental planning, ecosystem science, & ecosystem approaches for integrating environment & development. *Environmental Management, 17*(3), 289–303.

Somers, K., Baker, G., & Isbell, C. (1984). How to use the Delphi technique to forecast training needs. *Performance & Instruction Journal, 23*, 26–28.

Stabler, M. J. (Ed.) (1997). *Tourism & sustainability: Principles to practices*. New York: Cab International.

Stein, T. V., Clark, J. K., & Rickards, J. L. (2003). Assessing nature's role in ecotourism development in Florida: Perspectives of tourism professionals and government decision-makers. *Journal of Ecotourism, 2*(3), 155–172.

UK Department of Environment, Transport, & Regions. (2000). Public participation in making local environment decision. The Aarhus convention New Castle Workshop, London: UK DETR. Accessed June 2, 2003, from http://www.unece.org/env/pp

United Nations. (2001). *Managing sustainable tourism development: ESCAP tourism review*. No. 22. New York: UN.

United Nations (UN). (2002). *Studies in methods handbook of national accounting*. Series F No. 81: Use of macro account in policy analysis. New York: UN.

United Nations Conference on Environment & Development (UNCED). (1992). *Rio declaration on environment & development*. Rio de Janeiro: UNCED.

United Nations Economic Program (UNEP). (2001). *Soft mobility: Making tourism in Europe more sustainable*. Vienna: Tourism Focus, UNEP Tourism Program.

Veleva, V., & Ellenbecker, M. (2001). Indicators of sustainable production: A new tool for promoting sustainability business. *New Soultion, 11*(1), 101–120.

Wahab, S., & Pigram, J. J. (Eds.) (1997). *Tourism development & growth: The challenge of sustainability*. New York: Routledge.

Wallace, G. N., & Pierce, S. M. (1996). An evaluation of ecotourism in Amazonas, Brazil. *Annals of Tourism Research, 23*(4), 843–873.

Weaver, D. B., & Lawton, L. (1999). *Sustainable tourism: A critical analysis*. CRC for Sustainable Tourism Research Report Series, Research Report 1, Gold Coast, Australia: CRC Sustainable Tourism.

Williams, P. W., & Gill, A. (1994). Tourism carrying capacity management issues. In W. Theobald (Ed.), *Global tourism – The next decade* (pp. 235–246). Oxford: Butterworth Heinemann.

World Commission on Environment & Development (WCED). (1987). *Our common future*. Oxford: Oxford University Press.

World Conservation Strategy. (1980). *Secretariat/focal point*. Cambridge: World Conservation Union (IUCN), United Nation Economic Program (UNEP), World Wildlife Fund (WWF).

World Travel & Tourism Council (WTTC). (2004). *World travel & tourism: A world of opportunity*. The 2004 Travel & Tourism Economic Research. London: WTTC.

United Nation Conference on Environment and Development (UNICED). (1992). *Agenda 21*. Rio de Janeiro: UNCED.

Woudenberg, F. (1991). An evaluation of Delphi. *Technological Forecasting & Social Change, 40*, 131–150.

WTTC/WTO/Earth Council. (1995). *Towards environmentally sustainable development*. Madrid: World Tourism Organization.

Chapter 8
Island Awash – Sustainability Indicators and Social Complexity in the Caribbean

Sam Cole and Victoria Razak

Introduction

This chapter addresses the question of how far can tourism development proceed before the way of life enjoyed by residents of a highly successful small Caribbean island destination, Aruba, is irretrievably threatened by overdevelopment? Over the past decades, there has been recurrent concern that the island was approaching its carrying capacity. As the level of tourist activity has grown this, concern has become more acute, compounded by other related issues such as employment, immigration, and diminishing land resources. The current population of over 100,000 has almost doubled since the mid-1980s, mainly due to immigration. Driven by unprecedented growth in the 1980s and stop–go governmental policies since, the Minister of Tourism set up a National Tourism Council (NTC) to address the central question "how far and how fast" tourism should proceed in Aruba in order to confront issues of sustainability and carrying capacity. As long-time "friends of the Island," previously involved in preparing a macro-plan for independence from Holland in the mid-1980s (Cole, et al 1983), and authoring a variety of studies on the local economy and environment (Cole, 1997) and on festivals and culture (Razak, 1995, 2007a), the authors were invited to develop a framework for sustainable tourism development.

Aruba has among the highest population growth, density of population, and tourism of any small Caribbean destinations living on 180 km^2 and hosting nearly one million visitors a year (CBS, 2004). The challenges this poses for the sustainability of the natural environment, cultural heritage, and attractiveness to visitors are somewhat offset by the Island's relatively prosperous and secure society. Thus, the goal was to find an approach that maintained this enviable status within the constraints of its small island geography. The question asks what pace of tourism growth would supply Aruban's future needs, and would this level exceed the current

S. Cole (✉)
University of Buffalo, Buffalo, NY, USA
e-mail: samcole@buffalo.edu

M. Budruk, R. Phillips (eds.), *Quality-of-Life Community Indicators for Parks, Recreation and Tourism Management*, Social Indicators Research Series 43, DOI 10.1007/978-90-481-9861-0_8, © Springer Science+Business Media B.V. 2011

estimated carrying capacity? The Question is somewhat tautological in that "needs" are also central to our notion of carrying capacity. In terms of quality of life, the framework aimed to provide a steadily improved standard of living for all current Island residents, whether native or settled, and for successive generations, by redirecting tourism geographically and stylistically, thereby protecting and fostering the Island's heritage and prosperity.

Historically, Aruba's tourism has largely been oriented toward a recreational experience of "sun, sand, and sea" fostered by a combination of good climate and beaches, a professional tourism sector, public security, local entrepreneurship, a welcoming native ethos, and a hard-working migrant population. As a highly popular destination, the Island's success means that pressures from the industry for further development are intense. Today, tourism in Aruba, directly and indirectly accounts for around 60–65% of GDP and 65–70% of employment (Cole and Razak, 2004). Relative to the rest of the Caribbean, Aruba has the highest level of foreign investment and control (about 60% of rooms are marketed through international chains), with the advantages and difficulties that this poses for small island nations. While this relationship has enabled Aruba to develop a competitive industry, public and household income derived from tourism is less than some competitor islands. Moreover, opportunities for "destination branding" based on authentic cultural experience and heritage are neglected. Barely controlled new hotel and home construction has required a large influx of immigrant workers who often seek to settle; these new construction projects have led to the increased consumption of rapidly diminishing land resources. Thus, a combination of factors has led to a need to find a more "sustainable" path for tourism development. The challenge posed to the authors in 2002 was essentially to answer the question of how far and how fast should tourism proceed in Aruba, and how to position the goals and strategy for sustainable tourism.

In key respects, the question of "sustainability" is a minefield. Even if a philosophical foundation is agreed, questions of how to define, measure, monitor, and compare indicators remain intractable, most effort gauges the few items that in a more or less systematic fashion. Given a 1-year timeframe and dearth of relevant data, and recognizing that the work was part of a continuing island-wide discussion within a diversity of viewpoints, a pragmatic approach was inevitable managable given the complex structures, histories, and ambiguities of Aruban society. Indeed, it would be difficult to overcomplicate practically any issue in Aruba.

We begin by briefly summarizing the history of population and tourism growth in Aruba. Next, we explain how the ambiguities in the concept of tourism sustainability and the social complexity and policy setting have each influenced our approach. We then discuss how these issues were formally addressed in the proposed tourism framework using a combination of demographic, tourism footprint, and culture regions. We conclude with an appraisal of the mixed success of the study and the role played by our formulation of carrying capacity and analyses.

A Diverse and Dynamic Society

Aruba has experienced at least five epochs of globalization: pre-Colombian settlement of Amerindians; colonial occupation by the Spanish and later the Dutch;
transformation to a more multiethnic society via a large American oil refinery; a
prolonged negotiation for independence; and increasing globalization. Each era led
to a new wave of immigration so that the present-day society comprises peoples
from all parts of the world (the 2000 Census estimates approximately 100 nation-
alities). The overall growth of population and industry is shown in Fig. 8.1, and
especially the surges in population following the opening of the refinery in the late
1920s, and again during a rapid expansion of tourism after 1985. Today, Aruba comprises several distinct groups, including Native Arubans, Afro Arubans, and Émigré

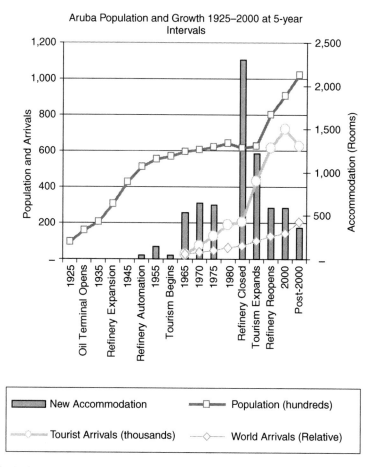

Fig. 8.1   Aruba tourism and population 1925–2005

Arubans; it is the cultural remnants of these settlers that provide the unique heritage upon which tourism – especially that aimed at providing a unique experience to visitors – must build.

Tourism Growth in Aruba

Tourism in Aruba began in the late 1950s when the government sought to offset layoffs as the oil refinery automated production. A 100-room hotel was opened in 1959 followed by two 200-room properties in the mid-1960s, then by 10 similar-size high- and low-rise properties by 1980, by which time Aruba had become a well-known and popular destination (ATA, 2003). Growth stalled in the early 1980s due to recession in Aruba's North American major markets. The Island was immediately confronted by the "final" shutdown of the refinery. The impact was devastating; some 30% of all jobs on the Island were lost, home values dropped dramatically, and the population fell as people sought refuge in the Netherlands and other destinations (Cole, 1997). With the added uncertainties of complete independence from Curacao and a looser relationship with Holland, tourism arrivals further declined.[1] To offset unemployment, the government accelerated the First Tourism Plan, already prepared in anticipation of independence (Spinrad, 1981; Sasaki, 1983). Unfortunately, the basis for estimating the number of hotel rooms

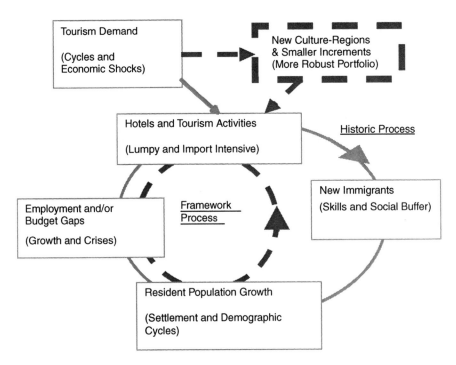

Fig. 8.2 Aruba growth dynamics and tourism strategy

required rested on a misunderstanding of tourism multipliers which resulted in considerable overbuilding.[2]

The construction and operation of the new hotels created a new "unanticipated" wave of immigration, mainly from the Spanish Caribbean and coastal Latin America where populations are ethnically quite similar to the Arubans, but not culturally or linguistically. Once again, the resident population felt swamped by a new wave of immigration. Thus, the underlying dynamic of development in Aruba is that, although the expansion of tourism is driving the economy, it is the growth of population that absorbs the available land and exacerbates issues of sustainability. In effect tourism – and especially the construction of large hotels – has created a "population" pump. This historic process is shown by the solid lines in Fig. 8.2. The countervailing process devised for the framework will be discussed later.

The accelerated growth of population and within-island migration have led to urban sprawl and increased attrition of the Island's natural landscape primarily adjacent to the Island's tourism corridor and the capital, Oranjestad, and encroaching onto the undeveloped North Shore of the Island. Thus, by 2002 these communities had accepted the need to rethink the distribution of tourism on the Island. In particular, the southern area around the oil refinery town of San Nicholas which has attractive beaches and a significant migrant population has almost no tourism development and could potentially support a variety of recreational activities.[3]

Conceptual Ambiguity

The most widely used heurist for considering a tourism destination's approach to capacity is the tourist area life cycle (TALC) devised by Butler (1980). This explains how, as a destination approaches some measure of "carrying capacity," growth slows and the destination stagnates or declines. There has been much critique and misinterpretation of the TALC. In the early 1980s, for example, both Spinrad (1981) questioned whether Aruba had not already reached its carrying capacity. Recently, Butler (2006) has observed that the TALC "was always envisaged as having several components and not as a single 'magic' number [which is] impractical to determine even in wilderness areas let alone in such a varied setting as a resort or destination." Moreover, he says, "Determining limits or maximum levels is difficult to achieve given the different interests in most destinations, and the range of viewpoints and timescales involved. The management and control of tourism is difficult because of the lack of unanimity or even agreement over the scale, direction, and control of development, accentuated by the variety of players and concerns of residents (reduction or loss of quality of life, rights), visitors (reduction of satisfaction, quality of experience, complaints), and policy makers (employment, acceptability, international relations)." [4]

Defining and measuring carrying capacity and sustainability are immensely complicated, arguments that it cannot be represented by a "magic number" are clearly correct. Nonetheless, the approach adopted for the framework for "measuring" carrying capacity or using it as an "indicator" was to provide a benchmark that would clarify community-wide understanding of alternative tourism strategies. In effect, the approach sought a lowest-common-denominator definition of carrying capacity that would provide a basis for discussion. Even in the absence of social complexity, any elaborate definition of carrying capacity would be justly vulnerable to philosophical and technical critique. On the other hand, we would argue, within the context of a complex island-wide negotiation, a plausible measure of capacity might serve as a useful rallying point and provide guidelines for assessing progress toward sustainability. It seemed more useful to provide a definition and measure that, once broadly agreed, could provide benchmarks for assessing future progress and directions. Moreover, if the Island chose to adopt this metric, it would likely tackle other specific environmental and cultural vulnerabilities beyond the tourism-related issues dealt with in the framework.

The thinking behind this approach draws on Cohen's (1985) discussion of the role of symbols in the construction of community. He suggests that "Symbols are effective because they are imprecise. Though obviously not content less, part of their meaning is 'subjective'. They are therefore the ideal media through which people can speak in a 'common language'... the symbolic repertoire of the community aggregates the individualities and other differences found within the community and provides a means for their expression, interpretation, and containment. It provides the range within which individuality is recognizable" (p. 21). Our concept of carrying capacity includes its symbolic significance, although unlike most symbols its indicators are precise (at least within the vagaries of the available data) but lose intent (and hence become ambiguous) without the wider context of the sustainability framework.

Saveriades (2000, p. 148) considers that most research on carrying capacity deals with the capacity for tourists or sociological capacity due to interaction between tourists and host population, but explains that such thresholds are perhaps the most difficult to evaluate (as opposed to environmental, cultural, and economic) since they rely entirely on value judgments. He argues that definitions of recreational carrying capacity share common elements. First is a biophysical component with some threshold or tolerance beyond which further exploitation imposes strains on the natural ecosystem. Second is a behavioral component in the quality of visitor experience when certain capacities, as perceived by the tourists themselves, are exceeded and a destination ceases to attract them (Mitchell, 1979; Wall, 1982, p. 191). Paralleling the first element is the sociological carrying capacity that is the host community's attitudes and tolerances that in turn set limits on tourist development (Allderedge, 1972; Doxey, 1975; Getz, 1987). Broadly, this last element determined the approach used for our study.

The indicator used to benchmark this tolerance derives from the concept of ecological footprint, first introduced by Wackernagel and Rees (1996). Formally, an ecological footprint is the land (and water) area of the planet or particular area required for the support of either humankind's current lifestyle or the consumption pattern of a particular population. Wackernagel and Rees (1996) showed that the lifestyle of a society could be translated into a land area equivalent – basically by converting the basket of goods consumed into its carbon content and then translating that into the land area needed to produce it. On that basis the USA is just about in balance, Iceland has a footprint 11 times its area, and roughly $2\frac{1}{2}$ worlds would be needed for everyone on the planet to live as Americans do today. Arubans (who do come close to that) have a footprint approximately 50 times the land area of the Island.[5] However, because practically everything is imported, most of that footprint is overseas, given the arid climate and limited opportunities for scale economies in production.

The value of the footprint concept is that it facilitates an overall quantitative measure of lifestyle, however crude, to be derived. As just indicated, one central question addressed by the concept is how dependent a community is upon others, for example, a small nonindustrial society on imports from overseas, a city on its hinterland, or even an individual on his/her community. With tourism, at least for the present study, the question is reversed, what is the footprint of tourists on the host destination and community? For present purposes, this is the amount of land needed to support each new tourist taking account of direct effects, accommodation, activities, and support structure, and the indirect effects on businesses, residences, and infrastructure. This facilitates qualified quantitative estimates for carrying capacity in relation to different strategies for tourism. While the footprint approach is quantitative and tangible, the benchmarks pose the more fundamental questions of "What is disappearing, what is cherished, and what is aspired to, in Aruban society?"

Social Complexity

A problem in applying any single concept of carrying capacity is that different segments of society adhere to, or dismiss, the competing paradigms (Barton, 1996). Indeed, in Aruba, discussion of any topic is complicated both by the fact that society comprises so many communities subdivided by nationality, language, country of origin, education, occupation, industry, kinship, political affiliation, and place of residence and the continuing turnover of population. Coalitions, identity, and consensus vary according to the issue at hand. Moreover, any attempt to impose a definitive prescription in Aruba from within or without is likely to be frustrated.

In terms of physical limits, the segment of society most concerned with the biophysical aspects of carrying capacity are Dutch trained civil servants and educators, and activists concerned with Aruba's remaining endangered species and natural regions. In terms of perceptual and psychological limits, the traditional population, living in the least penetrated parts of the Island (the rural Cunucu), ironically are more concerned about jobs and being crowded out by immigrants, while small businesses, many serving the tourist industry, are concerned with being crowded out by larger corporations. Table 8.1 shows the approximate correspondence between the broadly defined groups most concerned or impacted by tourism in Aruba and the variety of concerns implied by tourism sustainability. Given this, it may seem rather naive to adopt a single carrying capacity metric to assess the potential for

Table 8.1 Correspondence between sustainability concepts and tourism interests in Aruba

Concepts interests	Physical e.g., crowding	Perceptual e.g., friendliness	Economic e.g., costs, jobs, revenues	Institutional e.g., safety	Clinical e.g., cleanliness
Tourists "tropical island fantasy"	x	x		x	x
Tourist industry "high-end destination"		x	x	x	
Metropolitan Arubans' "suburban dream"	x		x		x
Expatriates "Tropical dream"	x		x		
Local commerce "business imperative"			x		
Educated elites "Dutch environmentalism"	x			x	
Native Arubans' "traditional caution"	x	x			
Migrants "survival imperative"			x		

Based on Cole (1993) and Wilkinson (2003); entries indicate areas of main concern.

new tourism. Weighed against this was the objective that, whatever their concerns, all Arubans could relate to the measure used, that it was immediately tangible, yet could also be incorporated into current trends and possible future directions.

These groups are by no means homogeneous with divisions across districts, barrios, kinship groups, and personalities. In addition there are contradictions such as the tension between tradition and modernity, most apparent among the native Arubans. With land and historic symbols disappearing and immigration threatening their culture, values, and language, the sense of loss becomes heightened, generating anxiety and a negative reaction to these changes. Confounding these issues further are generational changes, a pervasive rural traditional core, and modern ethos which support both the maintenance of a unique Aruban cultural heritage and the desire for a prosperous modern suburban lifestyle. Overall, the diversity of Aruba's population poses the seemingly contradictory questions of how to provide a distinctive new style of tourism that draws on particular components of her diverse heritage, but is not overly swamped by that diversity?

A diversity of views also was clear across the NTC whose membership included the Minister of Tourism and leaders from relevant government departments (finance, culture, social affairs, environment, and education) each of which independently control significant tourism-related resources and the Tourism Authority. The hotel association (controlled by the major hotels), the timeshare association, the chamber of commerce, and the banking community represented the private sector. Fortunately, and perhaps not surprisingly, given the motivation for the NTC, the core strategy of "Matching growth to Aruban needs" was at least given lip service as the starting point for a new strategy. That said, there were many nuances and agendas at play well beyond the comprehension of the present authors. One problem is that departments represent different interests but have overlapping jurisdictions relevant to tourism, so each can block another's proposals.

Since nobody stands against the idea of sustainability, the debate is rather about the trade-offs involved. For example, one influential position was that "sustainability is fine as long as it doesn't get in the way of business." Obviously, therefore, the framework as a whole had to tackle issues well beyond those of sustainability and carrying capacity, but these are of lesser concern for present purposes. The overall schematic for constructing the strategy is shown in Fig. 8.3. The left-hand elements indicate the public and private sector issues and concerns. These were tackled using a variety of economic models. The lower elements indicate the policy outcomes, including regional diversification and specific tourism products. The right-hand elements cover the aspects of most concern to the question of sustainability considered here – population growth and immigration, socio-economic needs, and the consequences for land use across the Island regions.

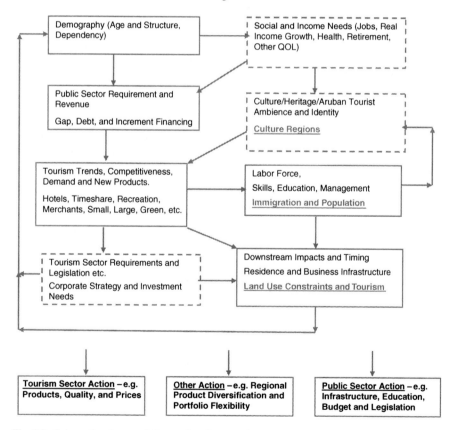

Fig. 8.3 Schematic of research for tourism framework

Overview of Planning Framework

The framework aimed to moderate the overall growth and fluctuations and regional imbalances within the socially defined carrying capacity. Given the small size of the Island in terms of geographic area, beaches and wilderness, and potential sites for tourism, there are upper limits on the number of hotels that may reasonably be constructed without overcrowding both tourists and residents. The most pressing constraint on tourism in Aruba however is determined to be the limited availability of residential land. Since the demand for residential development is driven primarily by the growth of tourism and tourism-related immigration, the recommendation is made that the total level of tourist accommodation should not rise to more than about 12,000 rooms over the next 40 years. The calculation behind this figure is explained below. The shift from a tourism based primarily on large properties operated by international chains with a high proportion of temporary migrant labor to smaller locally operated properties aimed to reduce the pumping effect noted earlier. The resulting greater stability of population would strengthen local cultures, training and improvements in local skills, both of which would raise incomes from tourism and maintain the dynamism of the industry.

The framework proposed that growth should be at a pace sufficient to meet the foreseeable needs of resident Arubans (native and migrant), provide full employment of the existing and future labor force, improve the situation of public finance, and achieve threshold scale economies. Calculations (described below) showed that this requires only between 50 and 150 rooms to be opened in any year. For a variety of reasons, stemming from the overall goals of the tourism framework, such as income capture, a more robust portfolio, and greater opportunities for local involvement, the recommendation was for a complement of smaller boutique-style hotels in less-developed regions of the Island (Cole & Razak, 2009). Any new larger properties, or expansion of existing hotels, should be carefully coordinated with other tourism activities, the labor market, infrastructure capacity, and public finances.

Slowing the pace of growth while continuing to improve living standards requires that the Island continually improves the scope and quality of the tourist experience. Regional diversification of tourism is a central aspect of the proposal and is designed to address several objectives simultaneously. Specifically, regional diversification would make better use of Aruba's tourism potential through a variety of new tourism products and projects. This would include the improvement and consolidation of existing tourism products within defined geographic areas or culture regions which would help to control the direct and indirect burden of tourism activities on already intensively developed areas and from overencroachment on the Island's natural environment (Razak, 2007b).

A primary goal of regional diversification was to draw upon the cultural diversity of the population in the development of new tourism products, while extending tourism activity into the less prosperous regions of the Island. This said, there were differences in the desire of the different communities across the Island to engage in additional tourism. For this reason, as well as the need to bring the clusters of activity up to a sufficient threshold (to take advantage of agglomeration and scale effects), the process was viewed as a succession of overlapping culture region TALC cycles "piggybacking" on existing mainstream Aruban hotel and timeshare-based tourism. The intended outcome of this approach on the growth of tourism by the major districts of Aruba (Noord, Santa Cruz, Oranjestad, and San Nicolas) is shown in Fig. 8.4. Residents of the most southerly town of San Nicolas (widely recognized as the "musical heart" of Aruba) have sought "low-key" tourism development for some years. With access to attractive beaches and access to the locally-known "wild" side of the Island, the area provides a viable location for clusters of smaller family and boutique hotels that together meet the economic requirements (in terms of scale, agglomeration, and so on) proposed for the next phase of tourism development in Aruba. In contrast, while the more traditional-minded people of the native Aruban community – clustered in the center of the Island – have the cultural base and physical attractions to develop a "folkloric" region, they have not participated greatly in tourism, and many appear less eager to do so in the immediate future. Thus, proposals for bringing a native Aruban flavor into mainstream tourism, together with educational heritage building projects for both tourists and locals alike, will be developed at a later phase when interest has increased.

The aim here is not to exclude the relatively indigenous population from tourism, in fact, quite the reverse since the ethos of this community gives the Island it deserved ambience of "One Happy Island." The goal is to bring the distinctive

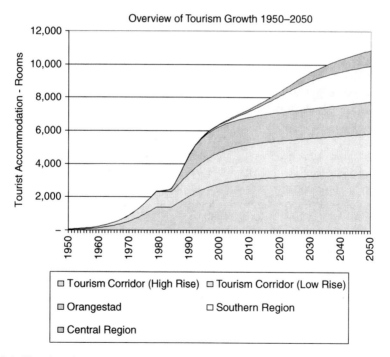

Fig. 8.4 Historic and proposed tourism growth by region

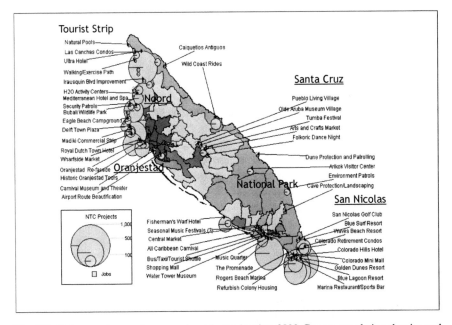

Fig. 8.5 Culture regions and proposed products showing 2000 Census population density and anticipated new employment by 2020

Aruban sensibility and flavor into the fabric of tourism rather than as a mere accessory to the dominant "sun, sand, and sea" atmosphere of Aruban tourism. Nor does this exclude this population from the income gained through tourism since many families are already engaged in tourism-related activities in the tourism corridor, in construction, small business, the public sector, and in hotel services.

The proposed levels of development for each culture region, in terms of new hotel accommodation, tourist activities, new dwellings, and related activities, were calculated (as described later) and projected some 40 years into the future. The purpose was to show that even with a growth that met the social and economic criteria, the Island would not have reached its carrying capacity. For each district also a of new tourism developments was suggested that were compatible with the styles of tourism proposed. Figure 8.5 maps the distribution of suggested projects over the next decade or so and their relative contribution to new employment opportunities in relation to the prevailing distribution of populations on the Island. Beyond this horizon, new projects consistent with the goals of the framework would arise.

Carrying Capacity and Sustainability

The approach to assessing sustainability possibilities is, in principle, rather straightforward. It consists of (a) an estimate of carrying capacity for new dwellings derived from available land area and total jobs per hotel room, (b) an estimate of future demand for homes (based on projections of labor supply) and preferred family structure, residence, and occupation, and (c) showing that (b) falls well below (a). It is complicated in practice by availability of data; other criteria for income growth and public sector finances; variability across tourism and hotel styles; the regional disposition of beachfront, cultural, and natural resources; and residual necessity for expatriate and immigrant workers. In addition to various demographic and economic information (derived from regular population and business surveys), the calculation rests on several "multipliers" such as the *total* jobs per new hotel room, the *total* land per new hotel room, and so on.[6] These multipliers were calculated from a social accounting matrix similar to that described in Cole (1997).

Knowing the number of dwellings required per new hotel room, the total land required per dwelling, and the amount of remaining available land, allows the carrying capacity for new dwellings and hence new hotel rooms to be calculated. Alternatively, knowing the "natural" growth rate of the current resident population, these same multipliers allow the annual requirement for new hotel rooms to be calculated. To avoid a lengthy recursive calculation, the procedure adopted was to set up a number of scenarios (in terms of hotels styles and locations) and calculate the required land distribution, and then compare this to the carrying capacity. The point of this calculation is not to claim some absolute limit to growth in Aruba, but rather it is to evaluate how imminent might be the limits to growth assuming that present styles of dwelling and access to nature remain a desirable aspect of Aruban life.[7] The aim therefore is to indicate the relationship between the present way of life enjoyed by Arubans and the territorial limits of the Island.

Tourism's Footprint

Given the amount of land remaining for development, and an estimate of a tourist's land-use footprint (based on Aruban lifestyle preferences and a given style of tourism development), it is possible to estimate the carrying capacity of the Island for new dwellings and hence new tourism development. As with other impacts, the footprint per room, beachfront per room, and other land-use "multipliers" vary across different styles of hotel, tourists, residents, and their homesteads. For example, low-rise beachfront luxury suites have a larger beachfront-to-tourist ratio than high-rise timeshare, ranch-style residences use more land per family than apartments. For Aruba, the average total land use, including all support activities – hotels, tourism, residence, and supporting business, and government – was found to be some six times that required for tourism accommodation and associated tourist activities (see Table 8.2). The carrying capacity of the remaining beachfront was estimated similarly. While a variety of technologies, changes in policy or tastes might extend carrying capacity considerably up to a "Caribbean Singapore," these would simply not fit within the image of Aruba that the Arubans articulate and the population would no longer be "Aruban."[8]

The contemporary Aruban aspiration for suburban single family homes with attached land, accessible to beaches and nature, pays homage to the traditional

Table 8.2 Hotel styles: tourism and total land requirement

Accommodation	Luxury 200–300 rooms	First class 200–300 rooms	Mid-price 200–300 rooms	Boutique under 100 rooms
Hotel footprint/room (m^2)	102	102	93	120
Tourists/room	1.9	1.9	1.9	1.9
Occupancy (%)	70	69	74	62
Visitor years	1.3	1.3	1.4	1.2
Other land use/tourist (m^2)	440	320	280	320
Tourism and support land area/new room (m^2)	542	422	373	440
Associated dwellings				
All workers/room	4.3	3.6	2.8	3.1
Workers/dwelling	1.5	1.5	1.5	1.5
Dwellings/room	2.9	2.4	1.9	2.1
Dwelling area (m^2)	500	500	500	500
Associated land (m^2)	1,250	1,250	1,250	1,250
Dwelling area/hotel room	3,584	2,996	2,368	2,621
Tourism and support land/new room (m^2)	542	422	373	440
Tourist use/total use (%)	13	12	14	14

Source: Cole and Razak, 2004

Aruban "cunucu" – a one-storey rural homestead and smallholding. About 85,000 new dwellings (over the present 29,000) could be built if all presently undeveloped land was used. However, since another goal for the Island is to protect unspoiled areas such as the National Park and the North Shore, the lower figure of 40,000 dwellings, corresponding to a population of up to 120,000 is appropriate. The average intensity of land use would then remain below 60% of the maximum level. With average levels of employment in hotels, average job multipliers, and number of residents per dwelling, this "target carrying capacity" translates to around 6,000 new hotel rooms.

The approach provided a means of estimating the upper limit on residential development based on the style of dwelling, the land needed for supporting infrastructure and business, the use of land for tourism, and the land that Arubans wish to remain untouched.

According to the Census 2000, there are 29,000 dwellings in Aruba. This represents an average of 586 buildings/km^2 within enumeration districts – those areas where live. This excludes undeveloped land in Census Districts (about 8%), and on the back of the Island, about 25% of the total area of Aruba. In 2000, about 14% of all land within census enumeration districts was taken for residential land (i.e., the domain upon which the building stands) although only about 3% of land is used for the actual building. In Oranjestad, the most developed district, approximately 24% of land is used, in Noord, Santa Cruz, and San Nicolas, the figures are 16, 11, and 13% respectively, excluding land taken for other uses such as roads, car parks, commercial centers, government offices, and tourism. The ratio of land used for business supporting these home varies so that in urban centers (e.g., Oranjestad and San Nicolas), commercial use tends to crowd out residential use. Thus, the highest intensity of land used for residential purposes is in the outer urban areas, at about 40% (Fig. 8.6). This pattern is fitted statistically (see Fig. 8.6) providing the relationship used to project the upper limit on land use.[9] There is one obvious outlier in this chart, a traditional "port of entry" for migrants in San Nicolas, to be discussed further below. This, and other locales familiar to all Arubans, provided a vehicle for emphasizing the implications of different tourism scenarios.

Several scenarios for the potential carrying capacity were developed. First, an "upper limit" scenario assumed that inhabited areas of the Island (outside of urban centers) approach a level of occupation implying that the pattern of suburbanization seen over the past decade eventually spread across the Island (shown by the dashed line in Fig. 8.6). Given these assumptions, it estimated that about 85,000 new dwelling units could be built if all presently undeveloped land was used. This shore-to-shore development would include about 31,000 dwellings along the North Shore. Another 10,000 would be in currently unpopulated areas within districts near the National Park that may ultimately prove unsuitable for residential development (such as rocky peaks, flood gullies, and salt flats). Assuming average levels of employment in hotels, average job multipliers, number of workers per dwelling, etc., up to 19,000 new hotel rooms could be accommodated, about three times the present number.

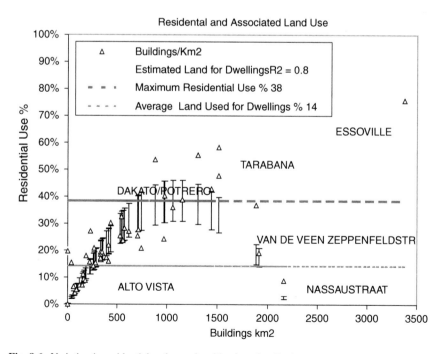

Fig. 8.6 Variation in residential and associated land use by district

A second scenario took account of new proposals by the government (drafted, but not then adopted) that the Island be subdivided according to five levels of land-use intensity: inner-urban, outer-urban, suburban, transitional zone, and protected areas (including the National Park). The principal implication of this proposal for the above calculation is the establishment of a less intensively developed transition zone (or buffer) between the suburbs and protected areas. Reducing development in this buffer zone and prohibiting further development along the North Shore would reduce the possible number of dwellings to about 43,000. Even in this scenario, it is likely also that residential densities would increase because of a reduction of terrain allocations and more widespread use of apartments, town houses, and condominiums. Since even this represents a total urbanization of the Island outside the protected and transition areas with densities similar to current inner suburbs, it is likely also that, should the present trend of urban sprawl continue unabated, land would be used more intensively through high-rise apartments or smaller dwellings, leading to other possible scenarios. Indeed, as a matter of policy the government has already reduced the average plot size allocated per dwelling.

Land Occupation with Tourism Strategy

A major question for the framework was whether this tourism capacity was sufficient to support the demographic, family income, and budgetary needs of the Island, and for how long? A two-generation horizon is not an abstraction for Aruba's family-oriented society. Thus, for this estimate, "Aruban" included everyone living on the Island at present and new immigrant settlement required to support development and their children and grandchildren.[10] Household income was to increase at a pace comparable to that expected for visitors to the Island. On average, this allowed for the construction of some 50–150 net additional rooms, equivalent to a cluster of small lodgings, a mid-sized hotel, or phased expansion of an existing property annually up to the 2040 time horizon.

Within this overall long-run target for development, tourism and population might grow and locate in a variety of ways to contain urban sprawl around the tourism corridor, and foster expansion at the southern end of the Island. This would still leave considerable flexibility to cope with new crises and contingencies. The sequential development of tourism across the selected Island regions was shown in Fig. 8.4. Over the entire period, the pace of development of the tourism sector is relatively smooth.[11] Local unemployment stabilizes at a relatively low level and the rotation of migration is steady. Moreover, the strategy is comparatively effective in terms of reducing imbalances in public sector finances and maintains average household income at or above the level offered by other strategies.[12]

The proposed strategy for tourism in Aruba is designed to slow residential growth significantly. With an overall population increase of 25% by 2045, for example, the average intensity of land use would remain below 60% of the maximum level calculated above. This population could be distributed in a variety of ways.

However, given the recommendation to consolidate tourism in the tourism corridor and Oranjestad, but foster significant expansion in San Nicolas, the greatest growth would be in this region. This area might also eventually be developed closer to capacity given its access to Arikok National Park and other attractive recreation areas. Conversely, since the central regions are mountainous and less accessible, the intensity of development would be somewhat less than other areas. The number of dwellings by district resulting from this proposed land allocation is illustrated in Fig. 8.7. The total number of hotel rooms envisaged in this scenario was about 11,000–12,000 rooms in total by 2045, well below the estimated carrying capacity.

Many other factors could be introduced into the above calculation, for example, the relationship between short- and long-stay migrants and differences between income and accommodation levels between immigrants from developed (mainly North America and Europe) and developing countries (mainly the Caribbean Basin). Census tracts with populations from poorer countries have a significantly higher density of homes, including the extreme instance remarked in Fig. 8.6.

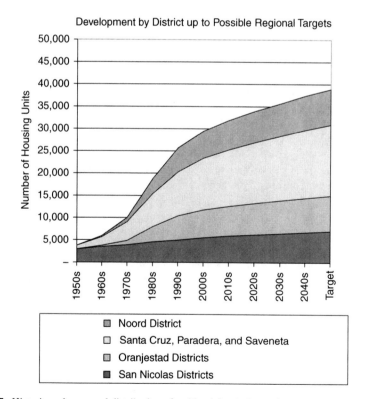

Fig. 8.7 Historic and proposed distribution of residential units by region

Aftermath and Gauging Progress

The goal of the framework was to provide a structure for discussion of tourism-related sustainability issues. In this respect the study has been relatively successful in that (according to observers) "it is cited in every meeting" and often referred to in the press more than 5 years since it was presented at the 2003 NTC. This longevity contrasts with most plans in Aruba, which rapidly become obsolete or overtaken in a capricious island society. While the issue of tourism-carrying capacity and sustainability features in the debate about Aruba's future, it is arguably primarily a vehicle for promoting, articulating, and debating other economic, social, and demographic concerns.

The simple sustainability measure and associated strategy addressed several issues: a high level of immigration and population growth, loss of nature around tourist strip where wealthier Arubans and expatriates live, suburbanization of the rural Cunucu regions occupied by the traditional Arubans, and nearer-to-home jobs for settled immigrants in San Nicholas. Insofar as a "magic number" approach worked as a vehicle for dialogue, it was because the concept could be used as an input to different interpretations and views. Insofar as the population of Aruba is lined up for and against the framework, it appears that the ambiguity of the sustainability indicator enables it to become a multifunctional rallying point, in that it may be used to reinforce several, even opposite, symbolic positions, just as Cohen (1985) argued for other symbols in the construction of community. For example, in favor is a loose coalition of small business, "nativistas," and environmentalists. The recommendations for specific proposals and activities (shown in Fig. 8.5) resonated with focus groups across the Island where the framework was presented. While those attending might be expected to be in general support of the framework and therefore not representative of their communities, they also are the "mavens" with influence in their respective communities.

In contrast, some government, larger business, and hotel chains and developers, including some members of the NTC, became lukewarm once the implications of the framework became clearer and new budgetary difficulties arose. Especially here, given the ambiguity of the indicator, the importance of retaining context becomes apparent. For example, one proposed private development treated the maximum capacity as if it was a target to be achieved within the next few years, with a single new resort development equivalent of 3,000 new rooms! (Cole & Razak, 2009). Similarly, on the political front, it appears preferable to retain the dependence on a migrant labor force, but to limit their settlement (by shortening the labor permits), and continuing the pattern that temporary migrants will continue to live in abnormally crowded conditions, whereas the framework assumed that all populations would eventually aspire to similar distribution of dwellings and services. Again, this position was argued from the settlement rates calculated for the study and the relationship between occupational density and migrants. In this sense, the carrying capacity analysis has backfired and may undermine its own objectives. In this use, the government has separated the indicator from its intended social context. Their alternative partially addresses the physical carrying capacity issue (through limiting settlement and reducing lot sizes) but, because it does not stabilize the population and associated environmental side effects, it also becomes far more difficult to raise educational and skill levels, or strengthen the Island's several identities, and hence raise the well-being and income to be gained from tourism. Over the long run, this limited view will exacerbate the problems of overdevelopment the north of the Island and increase social pressure for regional redistribution of development to the south.

Whatever the outcome, since the carrying capacity measure is rather simple, it is also straightforward to monitor, via the number of hotels, labor productivity, migration, land use, terrain size, and other indicators that are regularly collected. The proposals shown in Figs. 8.4 and 8.7 provide regional and temporal targets for each culture region and the Island as a whole. Through time we should expect the quality of data and analysis to improve and intermediate variables (such as demographic projections, productivity, dwelling footprints, and multipliers) to be revised. In addition, there are ad hoc qualitative measures of local sentiment in the many daily newspapers, Carnival and Calypso as forum for social commentary, or elections a manifestation of disappointment and disgust. While this metric does not cover key aspects of sustainability explicitly, if the Island ultimately embraces the framework, environment and heritage will be less impacted, and it will ensure that these will be addressed. The overall gauge would be that tourism and population growth slow down but the per capita income of Aruban's continues to rise and the pace of encroachment into undeveloped areas declines.

Notes

1. This limited independence called "Status Aparte" involved Aruba's withdrawal from the Federation of the Netherlands Antilles (Curacao, Bonaire, Saba, St. Maarten, and St. Eustatius) in favor of a one-on-one fiscal and governmental relationship with Holland.

2. The Island Department of Economic Affairs (UNDP/DECO, 1986) was working on a 1 room = 1 job basis (even less since the estimate was approximately 5 per room, 25 in other tourism, and 25 in new construction) (UNDP/DECO, Tables 14 and 15). Initially, a 1,500-room expansion was contemplated which – even with a more reasonable multiplier of 3–4 (approximately 1 job per room plus 1 job in other tourist activities, plus 2 more indirect jobs) – would have covered the 5,000 unemployed jobs. Unfortunately, the government also offered to guarantee all and any developers against loss if they built a new hotel in Aruba – an offer they could not refuse! The decline rapidly reversed and between 1985 and 2000, another 10 hotels with some 4,000 additional rooms constructed – about half dedicated to timeshare. Tourist arrivals boomed and markets expanded, but not enough. Two new hotel properties remained unfinished for several years and another was demolished. This "overshoot" led to the second major slowdown, and ultimately crippling financial obligation for the government, creating a motivation for further expansion!

3. At the time of independence, a well-crafted land-use plan also prepared based around the idea carrying capacity prepared by consultants from Curacao (PlanD2, 1981) that could have ameliorated some of the subsequent residential sprawl was not adopted, seemingly for reasons associated with the social complexity of the Island.

4. Wilkinson (1997, 2003), who has authored one of the more systematic studies of tourism planning in the Caribbean visiting Aruba and after reviewing several formal definitions of

carrying capacity, physical, perceptual, economic, and institutional, asserted that "there has not been any real advance in how to apply the concept in practice or in the understanding of what it means in principle." This is because, he observed, there is an inability to apply the concept because of "the conceptual and methodological complexity of the problem – there is a fundamental uncertainty about what is implied by carrying capacity, even when used in its most general sense."

5. European style requires approximately 5–10 ha per person. Aruba has a population of 100,000 and an area of 193 km^2. Global Footprint Network www.footprintnetwork.org/en/index.php/GFN/page/footprint_for_nations/.

6. The emphasis here is to avoid the mistake made in the recovery program implemented following the closing of the oil refinery – see earlier footnote.

7. Monaco, for example, the most densely populated nation state in the world, and also rich and attractive, has 30 times the current Aruba density.

8. This option was earlier proposed in the 1990s by a Government Minister who was summarily voted out of office.

9. This method was developed to circumvent the lack of current information on land use (e.g., residential, business, tourist, or multiple use), ownership (property land, long lease, short lease, etc.), etc.

10. In recent years, about half of all immigrants domicile on the Island with a high rate of intermarriage with Arubans, while temporary migrants stay an average of about 3 years (CBS, 2004).

11. In the short term, this requires coordination between public sector layoffs and speeded-up construction to offset the ripple effects of previous construction cycles. Fluctuations decrease as the increments in new accommodation are synchronized to labor force demands.

12. In the short run, this is due to the reduced labor costs. Over the longer run, public finances improve because the level of local ownership is higher and with this the possibilities for more successful tax collection.

References

Allderedge, R. B. (1972). Some capacity theory for parks and recreational areas. Washington, DC: USDI National Park Service Reprint.

ATA (2003). The history of tourism in Aruba mimeo. Oranjestad: Aruba Tourism Authority.

Barton, M. (1996). The social anthropology of complex societies. London: Tavistok.

Butler, R. (1980). The concept of a tourism area life cycle of evolution: Implications for management of resources. *Canadian Geographer, 24*(1), 5–12.

Butler, R. (2006). Recycling the Cycle: The TALC and Sustainable Tourism in the 21st Century. University of Strathclyde.www.dit.ie/media/documents/research/tourismresearchcentre

CBS (2004). Double or quits: Migration to Aruba 1993–2003. Oranjestad: Central Bureau of Statistics.

CBS. Annual tourism profile of the social economics statistics division of the Central Bureau of Statistics, 2001–2007, Oranjestad, Aruba.

Cohen, A. (1985). The symbolic construction of community. London: Routledge.

Cole, S. (1993). Cultural accounting in small economies. *Regional Studies, 27*(2), 121–136.

Cole, S. (1997). Economic cultures and ecology in a small Caribbean Island. In J. van den Berg & J. van der Straaten (Eds.), Economy and ecosystem in change: Analytic and historical approaches. London: Edward Elgar.

Cole, S. Opdam, H., Ahrends, B., Zambrano, R., & Razak, V. (1983). Aruba: Planning for independence. Oranjestad: Institute for Applied Economic Research/Department of Economic Affairs.

Cole, S., & Razak, V. (2004). A Framework for sustainable tourism in Aruba, Prepared for The National Tourism Council of Aruba and The Minister of Tourism and Transportation, December.

Cole, S., & Razak, V. (2009). How Far and How Fast? Population, culture, and carrying capacity in Aruba. *Futures*; June

Doxey, G. (1975). A causation theory of visitor-resident irritants: Methodology and research inferences. Sixth annual conference proceedings of the travel research association. pp. 195–198. UT:TTRA. Financial Mirror. 1998, November 25. Better tourism, hi-tech to boost

Getz, D. (1987). Capacity to absorb tourism: Concepts and implications for strategic planning. *Annals of Tourism Research*, *102*, 239–261.

Mitchell, B. (1979). Geography and resource analysis (pp. 176–200). New York: Longman.

PlanD2. 1981. Aruba Spatial Development Plan 1981–1990. Fudashion Desaroyo Planea, Oranjestad, Aruba.

Razak, V. (1995). Culture under construction: The future of Aruban identity. *Futures, the Journal of Policy, Planning and Futures Studies*, *274*, 447–461.

Razak, V. (2007a). Carnival in Aruba: A feast of yourself. In G. L. Green & P. W. Scher (Eds.), Trinidad carnival: The cultural politics of a transnational festival. Bloomington, IN: Indiana University Press.

Razak, V. (2007b). From culture areas to ethnoscapes: An application to tourism development. *Journal of Regional Policy*, *373*, 199–212.

Sasaki Associates (1983). Aruba tourism development plan. SA 1141. January. Watertown, MA: Sasaki Associates.

Saveriades, A. (2000). Establishing the social tourism carrying capacity for the tourist resorts of the east coast of the republic of Cyprus. *Tourism Management*, *21*, 147–156.

Spinrad, B. (1981, March). Aruba tourism development plan (Vol. 1). Miami, FL: The Tourism Industry, International Development Advisory Services.

UNDP/DECO. (1986). Plan national de Desarrollo. Department of Economic Affairs. April 1986.

Wackernagel, M., & Rees, W. (1996). Our ecological footprint: Reducing human impact on earth. Gabriola Island, BC: New Society Publishers.

Wall, G. (1982). Cycles and capacity: Incipient theory of conceptual contradictions. *Tourism Management, 3(3)*, 188–192.

Wilkinson, P. (1987). Tourism in the small island states of the Commonwealth Antilles. Caribbean Studies Association, London, England, July 29–31.

Wilkinson, P. (1997). Tourism policy and planning: Case studies from the commonwealth Caribbean. York: Cognizant Communications.

Wilkinson, P. (2003). Carrying capacity (CC), tourism, and islands microstates University of Toronto, Aruba National Tourism Conference 2003.

Chapter 9
Tourism-Specific Quality-of-Life Index: The Budapest Model

László Puczkó and Melanie Smith

Introduction

Tourism is a complex industry. It provides employment opportunities and tax revenues and supports economic diversity. It has very different impacts, both positive and negative, or even mixed ones. However, from a national, regional or local planning point of view, tourism should support the development of the quality of life of citizens too. If the implications of tourism have mainly negative impacts on the local community and environment, citizens will not support tourism and will not welcome visitors. This could mean a sad end to a destination.

The Budapest model introduces an integrative approach to quality of life (QOL) studies, applying QOL theory and practice to the field of tourism. The model was developed by a multi-disciplinary team of academics and practitioners, who combined theory and research to develop a TQOL (tourism quality of life) Index and model which can be used to measure the quality of life of both residents and tourists in a destination. This includes investigating five identified TQOL domains, which are attitudes towards travelling (ATA); motivations of the visitor (MV); qualities of the trip (QT); characteristics of the destination (CD); and impacts of tourism (IT). The preparation of the model was prompted by the Hungarian Tourism Development Strategy (NTDS 2005–2013). One of the key objectives of the strategy was to enhance quality of life through tourism. This objective was based on the hypothesis that tourism could make a complex contribution to improving the quality of life of both local citizens and visitors.

As part of the implementation process of the NTDS, in 2006, some preliminary actions were taken in order to provide a basis for the development of applicable methodology and tools. The first outcome was the preparation of a document with the title 'Foundations for the Preparation of the Tourism Related Quality

L. Puczkó (✉)
Corvinus University, Budapest, Hungary
e-mail: Lpuczko@xellum.hu

M. Budruk, R. Phillips (eds.), *Quality-of-Life Community Indicators for Parks,*
Recreation and Tourism Management, Social Indicators Research Series 43,
DOI 10.1007/978-90-481-9861-0_9, © Springer Science+Business Media B.V. 2011

Fig. 9.1 Main domains of
quality of life (after Rahman
et al., 2005)

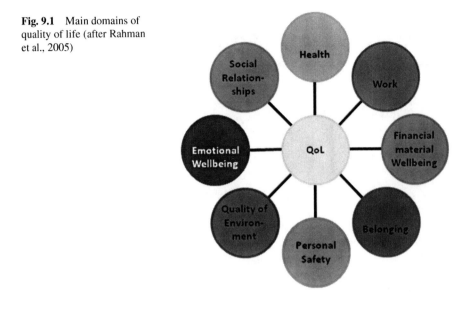

of Life Index' (Kovács, 2006). From the many definitions and models, Rahman, Mittelhammer, and Wandschneider's (2005) approach was selected. It was anticipated that Rahman's model can serve as a sound basis for the preparation of the Tourism-Specific Quality of Life Model and Index (TQOL-I). According to Rahman's approach, there are eight factors with a special influence on QOL. It is anticipated that all of these eight domains, directly or indirectly, have some kind of relationship to and with tourism. The model was first piloted in Hungary in three tourism destinations, and then refined following an international testing phase in ten other countries. Participants in the survey and other QOL experts were then invited to a roundtable discussion meeting in 2008 to discuss the strengths, weaknesses and future recommendations for the model (Fig. 9.1).

Overall, this chapter will introduce the theoretical and practical foundations of the Budapest model, as well as presenting the most important findings. It will also provide some reflections on the future refinement of the approach.

Tourism and QOL

Many researchers have been debating the meaning of quality of life since the 1960s, and there have been several indices developed to measure it. It is clearly difficult to differentiate between such terms as 'well-being', 'welfare', 'happiness', and 'quality of life'. What is clear, however, is that quality of life is a complex, multi-dimensional phenomenon, which incorporates not only elements relating to standard of living, but many other elements. For example, Fekete (2006) suggests that quality of life combines the material elements of standard of living and welfare with the intangible dimensions of well-being. Quality-of-life indices should therefore not only measure objective, quantitative economic and financial factors, but subjective, qualitative social and perceptual ones too. Fekete (2006) suggests including the following factors: sustainable development, human development, livability, social inclusion, exclusion, cohesion and capital. Campbell, Converse and Rogers (1976) suggested health, private life, material well-being and productivity be included. Kim (2002) enlisted and analysed most of the available theory and models. Flanagan (1978) and Krupinski (1980) added to these domains emotional well-being, and Cummins et al. (1994) and Cummins (1997) added community and security. Indeed, it seems that the factors identified by Cummins (1997) are the most relevant to quality-of-life research. These include material well-being, often known as standard of living (measured by income), living conditions or circumstances, type of housing, social and material status. Community well-being is also important, including the local council and other services, business services, environmental quality, aesthetics, etc. (Sirgy & Cornwell, 2001).

Emotional well-being may refer to free time, religion or spirituality (after Fekete, 2006), ethics, morals, recreation and hobbies. Here we could add the domain of travel, including the anticipation of a trip, the trip itself, and post-trip satisfaction, which can all contribute to well-being or happiness (Neal, Uysal, & Sirgy, 1995, 1999). The domains of health and security are also very important to quality of life research, both on an individual and a collective or community level. Of course, subjective perceptions of these domains may vary from 'objective' measures. On a personal level, love, friendship, family relationships, status, and authority can also play an important role. Overall, many researchers have concluded that the following three domains are the most important for quality of life: health, standard of living and well-being.

Experts and researchers representing various fields of study, e.g. sociology, tourism, geography and economics, were involved in the preparation of the Budapest model. The research team overviewed the available and relevant references and resources in QOL and tourism and they had to conclude that relatively little research and modelling had so far been done in tourism. The likely role of tourism in QOL was not really highlighted either in general QOL or in social indicator studies; however, some research had been undertaken on well-being. Only a few tourism or travel sources were identified dealing directly, or more likely indirectly, with QOL (e.g. the works of Campbell et al., 1976, Cummins, 1997, Krupinski, 1980, and Neal et al., 1995, 1999). The most relevant of all was the research of Sirgy and Su (2000) or Kim (2002). They assumed that satisfaction with travelling may be related to satisfaction with life and positive changes of QOL.

Summarising the available research findings and approaches we find that

- Perdue, Long and Gustke (1991) analysed the relationship between level and state of tourism and QOL of local citizens. They took the following factors into consideration:

- ○ population,
- ○ economy (incomes, revenues),
- ○ education,
- ○ health services,
- ○ well-being, and
- ○ delinquency.

They concluded that tourism has an influence on migration, type of employment, cost and quality of education and health services provision.

- ■ In their study of objective indicators of the impacts of rural tourism, Crotts and Holland (1993) summarised that tourism has a positive influence on the QOL of rural populations. Among the positive impacts they identified were income, health, recreation and personal services, whereas they found tourism had a negative impact on poverty.
- ■ Johan (2004) identified four factors: socio-cultural, health-related, safety and environmental factors. The Delphi research aimed at the formulation of Tourism-Specific QOL Index in holistic tourism.
- ■ The research by Perdue, Long and Kang (1999) about (casino) gambling tourism and QOL analysed six factors: safety, social changes, involvement in community matters, traffic congestion and transportation, impacts of local politics, and changes in employment opportunities.
- ■ Jeffres and Dobos (2001) analysed how the media, leisure opportunities, and the related perceptions may influence QOL statements.
- ■ Olfert (2003) noted that the type and stimuli of residence can articulate the frequency of travelling and type and quality of destination visited.
- ■ Among others, Estes and Henderson (2003) also refer to the works of Csikszentmihályi, who was one of the first to analyse the relationships between certain activities and the experiences derived from those. These experiences were found to be linked to movement, the power of the new, the opportunity of getting involved, exploration, creativity and spending time valuably. Many of these factors somehow related to tourism.

Non-QOL-specific, especially marketing, management and planning tourism research also influenced the thinking of the research team. Most of the findings, however, had to be translated to QOL terms.

The extent to which tourism contributes to the quality of life of tourists, local residents, and destinations will be affected by different factors, for example, the importance of free time and holiday allowance within a society (Richards, 1999); the number and characteristics of the tourists visiting a destination (Rátz & Puczkó, 2002); the motivations and subsequent activities of tourists in a destination (Ajzen & Fishbein, 1977; Iso-Ahola, 1982; Gnoth, 1997); tourist satisfaction and perceptions (Neal et al., 1999); the characteristics of the local environment (Uysal &

Noe, 2003) and the communities living there, or the instrumental performance and expressive attributes of a tourist destination (Swan & Combs, 1976; Uysal & Noe, 2003).

Besides the limited availability of previous works, the research had to take the following issues into consideration:

- Many studies faced the problem that they could register only the results of some changes (e.g. Scottish Executive identified the relationships of cultural services to QOL), but did not map either the process or the drivers behind the impacts.
- Satisfaction may not automatically mean improvement in QOL, since satisfaction is one, but far from being the only element of well-being or QOL.
- The relationship and exact meaning of QOL and related phenomena, such as happiness or well-being, are not yet defined or very clear. This is still the case both for researchers and especially for the general public.

To add a little to the complicated background to QOL and tourism, besides tourists, the local population needed to be considered. According to QOL and social indicator studies, the residents' QOL depends on internal and external factors. The internal factors are (Andrew & Withney, 1976; Cicerchia, 1996; Cummins, 1996; Sirgy, 2001):

- Material well-being: subsistence, income, employment
- Communal well-being: people in the community; public space, life and services
- Emotional well-being: factors related to free time or spiritual well-being
- Health and safety-related well-being: both on personal and community level
- Factors of QOL: satisfaction with life in general

We could see before that the satisfaction with life can be influenced by the impacts tourism may have on locals. These impacts can be and are different according to the stage and level of tourism development. These economic, social, cultural and environmental impacts are seen as external factors to the perception of the local community (e.g. Ap, 1990; Lankford & Howard, 1994 or Liu & Var, 1986).

It is still surprising that travel and tourism have featured so rarely in quality of life research, especially given the global reach and appeal of tourism. However, it is acknowledged that tourism is only one of the factors which can have an impact on QOL, (e.g. through the economic and social development of a destination, the life and life conditions of those directly or indirectly working in tourism, or those who are resident in a tourism destination, not to mention those who are actually travelling). Tourism can influence the level of satisfaction with life and based on personal experiences as a tourist the reactions to changes in QOL at home. Sirgy (2008) suggests that life satisfaction can be increased by engaging in life experiences such as travel and tourism events, which can produce a positive affect in important life domains and allow that positive affect to spill over into one's overall life.

Definitions and Hypothesis

According to the objectives of this research, the main task was to develop a model, which through its complex measures analyses the person, i.e. the 'homo turisticus', the main actor in leisure and work-motivated mobility. Based on this framework, two hypotheses were formulated:

- Tourism does have some links to general QOL
- Tourism influences QOL

The research team selected Meeberg's (1993:37) definition: 'Quality of life is a feeling of overall life satisfaction, as determined by the mentally alert individual whose life is being evaluated', based on which the working Tourism-Specific Quality-of-Life Model (TQOL) definition could be formulated:

"Quality of life is a feeling of overall life satisfaction, as determined by the mentally alert individual whose life is being evaluated. In the formulation of the individual's opinion, which is fundamentally based on subjective factors, tourism can play a role".

Based on the findings of motivation, satisfaction and impact models and studies, and incorporating the general QOL factors, the formulation of the TQOL model was characterised by the following questions:

- What is important in life? What does make one happy? What is the role of tourism and travelling in that?
- What do people think of tourism as an activity?
- Why do people travel?
- What are the experiences during travel? What is the role of satisfaction?
- How would they describe the destination they visited?
- From Rahman's eight domains, which ones are related to tourism?
- How can the domains be measured? What factors are to be analysed?

The Budapest model incorporates the qualities of both objective (e.g. Diener & Suh, 1997) and subjective (Andrew & Withney, 1976) indicators, as most researchers agree that there are both objective and subjective indicators for measuring quality of life. These have been used in various models, for example

- The Scandinavian model which considers objective or quantifiable measures like standard of living and living conditions (Erikson, 1993). One of the key benefits of objective measures, e.g. tourism revenue per capita, is that they are relatively easy to quantify (Diener & Suh, 1997)
- The American model which examines subjective indicators (Campbell et al., 1976) that are rooted in psychology. Subjective indicators take those parameters into consideration that are important to the individual (Andrew & Withney, 1976); therefore, well-being can be measured based on level of satisfaction, for example

The recommended *domains* for TQOL are as follows:

1. *General attitudes towards travelling*:

   o Opinions and attitudes even of those who do or did not travel
   o Impacts of tourism and travelling on personal welfare and well-being
   o Impacts of travelling on satisfaction with life

2. *Motivation of the traveller/visitor*:

   o The influence travelling may have on happiness, satisfaction and well-being, depending on the type of trip
   o Trips are initiated based on different expectations and motivations; therefore, the impact on satisfaction can be very different depending on what type of trip one went on (e.g. health trips vs. city visits)

3. *Qualities of the trip and stay*:

   o Qualities of the movement itself (e.g. length, distance) and the satisfaction of which can influence well-being
   o Satisfaction with various qualities of the stay can influence well-being

4. *Qualities of the destination*:

   o Availability of (tourist) services
   o Quality of (tourist) services

5. *Impacts of tourism*:

   o Impacts of tourism (and not only tourists) on the local environment
   o Perceptions of the impacts of tourism

The new model aimed to be a tourism-specific one. This means that all general QOL domains had to be analysed from the tourism point of view. Or rather, the general QOL domains and factors were used, but all were applied to tourism.

Based on the Rahman et al. (2005) model, TQOL identified five domains:

1. Attitudes towards travelling (ATA)
2. Motivations of the visitor (MV)
3. Qualities of the trip (QT)
4. Characteristics of the destination (CD)
5. Impacts of tourism (IT) (Fig. 9.2)

According to the combined approach, most of the domains have both subjective and objective factors. Since the TQOL model is expected to be a tool for planning and management, the factors were allocated to fit mainly destination-level needs.

The subjective factors were analysed by using a questionnaire. The questionnaire aiming at the identification of subjective information had four main parts:

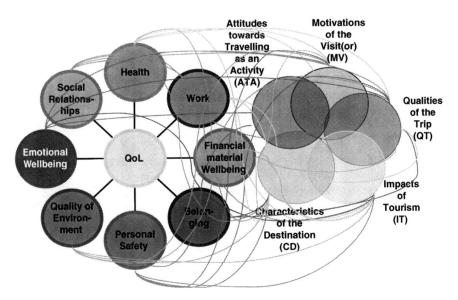

Fig. 9.2   The relationships between general and tourism-specific QOL domains

- Questions related to general QOL (subjective parameters – 46 questions), e.g.

  o Level of happiness
  o Statements about general QOL factors (e.g. life, family, work, health and travel)
  o Level of satisfaction with QOL factors/domains (e.g. life, family, work, health and environment)

- Questions related to general QOL (objective parameters – 11 questions), e.g.

  o Number of trips
  o Length of stay
  o Types of activities pursued

- Questions about the five tourism-specific QOL domains (97 questions), e.g.

  o The Attitudes towards travelling as an activity – ATA domain
  o Tourist/travel motivations – TM domain
  o The qualities of the trip – QT domain
  o Characteristics of the destination – CD domain
  o Impact of tourism – IT domain

- Demography questions (21 questions), e.g.

  o Marital and employment status
  o Educational qualifications
  o Type of job (related to tourism?)

While the subjective factors can be analysed using the questionnaire, objective factors are constructed by collecting numerical, mainly 'per capita' data. The objective factors, which were applied according to the recommendations of the so-called Scandinavian QOL model as mentioned earlier, accumulated numerical information. These data can indicate the role and type of tourism in the area where TQOL is measured:

- Attitudes towards travelling:

  o Number of trips per capita (domestic and foreign trips)
  o Per capita spending during trips
  o Number of visitors at attractions

- Qualities of the trip:

  o Number of people travelling together
  o Frequency of visits
  o Length of the trip
  o Demographics of the visitors (age, occupation, etc.)

- Characteristics of the destination:

  o Number of bedspaces per capita

- Impacts of tourism:

  o Number of (domestic and foreign) visitors per capita
  o Number of guest nights per capita
  o Number of employees in tourism
  o Average length of stay
  o Balance of tourism per capita

In Table 9.1, the interlinkages between the domains and factors are indicated.

Initial Test of the Budapest Model

Two research methods were applied: focus groups and field surveys. It has to be noted that not only during the pilot data collection but in the literature review too the team had to face the interchangeable use of the two terms: quality of life and happiness. For the sake of the pilot study, although we do not consider the two terms to be exactly the same (i.e. happiness is seen as more short-lived than QOL and represents 'highs' or peaks of positive emotions), we used them synonymously. This was mainly because respondents frequently used the term 'happiness' in preference to 'quality of life'.

Qualitative Research

The hypothesis of the qualitative test was as follows: 'It is anticipated that tourism plays a relatively significant role in how people evaluate their quality of life'. Two focus groups were organised. During the discussions the following questions were discussed in great detail:

- Scope, meaning and elements of quality of life and happiness
- The role of travel and tourism in quality of life and happiness (including both foreign and domestic trips)
- Does it have any impact on residents' quality of life and happiness if visitors come to where they live?
- Discussion about the five major domains (shown in Table 9.1)

Table 9.1 Relationship between TQOL domains and objective/subjective factors

	Objective	Subjective	Local residents	Visitors
Attitudes towards travelling	√	√	√	√
Motivations of the visitors		√	(√)	√
Qualities of the trip	√	√	(√)	√
Characteristics of the destination	√	√	√	√
Impacts of tourism	√	√	√	√

Quantitative Surveys

For the samples of local citizens, the team applied a stratified selecting procedure, while the samples of visitors were based on random sampling. Not only were the questionnaires tested, but the applicable scales too. In every location a different 5- and 7-point Likert scale was used (between –2 and 2; between 1 and 5; and between 1 and 7).

Not very surprisingly, one of the most difficult and complicated tasks for the participants was to define the concept and meaning of 'quality of life'. It was quite clear that 'quality of life' seemed to be a non-definable phenomenon that can easily be misinterpreted as standard of living. Especially for younger participants, quality of life seemed something that would concern them later during their life. They were more concerned about their standard of living, i.e. the material aspects of living. Interestingly, to many, travel appeared to be part of the material set, i.e. something that is part of a certain standard of living ('And, where will you travel this summer?'). It was concluded that standard of living is much easier to define and discuss since it depends on a smaller number of factors than quality of life and there is a measure that seems objective enough to quantify it, i.e. money.

According to the opinions of the participants, quality of life comprises the following:

- Financial security
- Family, love and relationships (both to give and receive)
- Tranquillity, peace
- Health status

As the concept and meaning of quality of life turned out to be too complex and vague, some participants suggested replacing it with harmony (terms such as 'emotional well-being' were far beyond the mindset of most participants).

The terminology used by participants for describing travelling included both travel and tourism. It was also highlighted that trips are characterised by the most memorable experiences gained during that trip. This would include both favourable and unfavourable experiences. During the discussions participants could list and discuss many likely factors that may determine quality of life. Since travel did not come up spontaneously as an influencing factor, moderators had to mention the term 'travel' as a possible factor. Interestingly, as soon as the option of travelling was raised, almost all participants agreed that some form of travelling would really be an important factor. Some relevant factors, however, were specifically highlighted:

- Although VFR visiting friends and relatives (VFR) trips were not considered as part of travelling, visiting friends and relatives still became an important factor of improving or maintaining relationships (which is a significant element of quality of life)
- Travelling for health reasons (medical, wellness or prevention)
- Impacts of tourism on employment opportunities

Results – Quantitative Method

The standard questionnaires included questions about both general and tourism-specific issues that may affect quality of life. The results confirmed that:

- Happiness in general
  - o If a local person saw him/herself happy that was mainly determined by family-related qualities
  - o Visitors defined themselves happier than local citizens of the destination they visited. This should make us think that travelling could make people happy (or happier)
- Role of travelling in happy living – this role was definitely identifiable and quantifiable, but it did not reach the highs of family, love or health.

o Approximately 20–25% of both visitors and local citizens rated tourism as a factor of happiness
o Interestingly, tourists thought that happier people were more likely to travel, and not that travelling makes someone happy
o While for local people, trekking, reading or spending time with kids/grandchildren meant the main forms of relaxation, for visitors, on the contrary, travelling was the ultimate form

The meaning of standard of living and quality of life followed similar patterns to that of the qualitative method: quality of life depends on level of satisfaction, happiness and health as well as standard of living and material goods.

The different Likert scales did not have an impact on the results but proved that scaling methodology can influence respondents' reactions. This was especially the case for the 5-point Likert scale ranging from −2 to +2. Detailed analysis of the data suggested that the results became biased towards the positive end of the scale. The motive behind that was the intention to avoid selecting 0 as the middle, i.e. the neutral element of the scale.

Many questions were devoted to the main determinant domains. We had to conclude that all domains and factors can be approved since none of those became neutral in the data analysis.

We could conclude that

■ There were no questions in the questionnaire which could be identified as unnecessary
■ The hypothesis can be accepted, i.e. travelling and tourism do play an important role in how people feel
■ According to the responses, the most important impact of travelling was that it could enrich life and make people happier

Tourism and travelling in general did not tend to come up top-of-mind as significant factors. Referring to specific trips, however, most of the respondents agreed that it did improve their quality of life and made them happier.

International Phase of the Research

Following the successful piloting of the survey research in Hungary in 2007 in Eger, Siófok and Gyula, the Ministry of Local Government (Hungary) decided to pilot the TQOL survey further and organised an international testing project. Members of international research and education networks (e.g. TTRA, ATLAS and AIEST) were invited to take part and international organisations (e.g. UNWTO and EU) were also informed about the developments. The model was tested in a further ten countries in 2007 and 2008 using the questionnaires mentioned earlier.

Each partner followed a similar sample size and sampling method in capturing data:

■ 50 residents and 50 visitors were interviewed, and
■ random sampling was applied.

Table 9.2 Countries
and destinations involved in
TQOL research

Country	Destination/location
Brazil	Sao Paulo
Czech Republic	Prague
Finland	Porvoo
Malaysia	Kota Kinabalu, Sabah
Portugal	Algarve
Russia	Moscow
Slovakia	Bratislava
South Africa	Potchefstroom
Holland	Harkstede
United Kingdom	Sunderland
Hungary	Eger
Hungary	Gyula
Hungary	Siófok

Data collection did not and could not have aimed at becoming entirely representative. The purpose was the analysis and testing of the methodology, i.e. the relationship between subjective and objective factors and to provide a wider basis for the index definition. Based on data collection by the partners listed in Table 9.2, a total of 960 questionnaires were completed and tabulated, and then processed in SPSS.

As a representation of how the collected data can be analysed, several figures are introduced here by way of examples. Figure 9.3 indicates that the general subjective QOL was similar for both local residents (res) and tourists (t) at most locations. Some exceptions could be identified, e.g. in the case of Harkstede (The Netherlands)

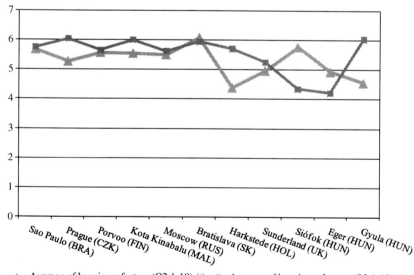

—▲— Average of happiness factors (Q2.1-10) (t)   —■— Average of happiness factors (Q2.1-10) (res)

Fig. 9.3 Average happiness factors

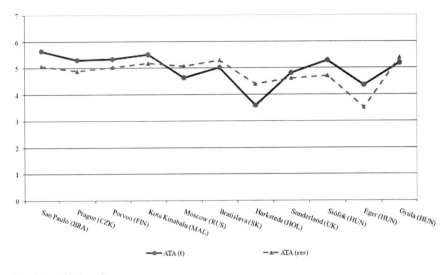

Fig. 9.4 ATA domain

or Siófok (Hungary). This information is especially telling, if we later look at the
TQOL Index results (Figs. 9.7 and 9.8).

Figure 9.4 indicates the respondents' views regarding tourism as an activity. It is
quite interesting that at many locations the ATA domain shows higher figures for the
residents than for tourists. This is somewhat surprising, since it is generally assumed
that visitors during their holiday are happier than not being on holiday.

The characteristics of the destination (CD) domain (Fig. 9.5) indicates that something may be not optimal for Harkstede as a tourist destination nor a place of
residence, as the relatively negative opinion of visitors echoed the locals' responses.

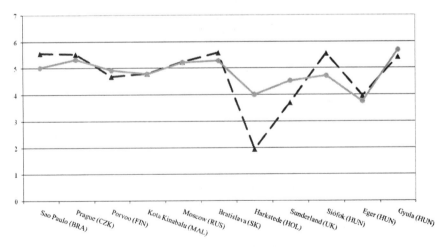

Fig. 9.5 CD domain

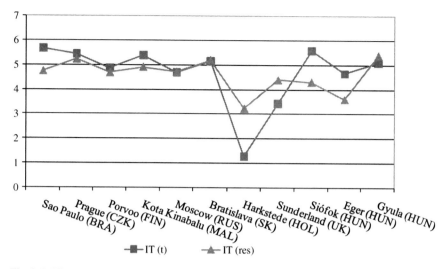

Fig. 9.6 IT domain

The same is true of Eger in Hungary, although to a lesser extent. This is a good example of cases where further analysis of the data collected could provide a clearer insight into these negative perceptions.

In Fig. 9.6, the results of the Impacts of Tourism domain can be found. Learning from the detailed analysis of the data we can add that at those locations where the results were low, the respondents added that impacts of tourism on the natural environment in particular contributed negatively to their QOL.

At the same time that the research was taking place, a specialist roundtable meeting was held in Budapest in 2008, where the representatives of research agencies and other organisations engaged or interested in the subject matter discussed the following:

- Role of tourism and related sectors in the quality of life of residents and tourists
- Experiences and findings gained from the collection of data
- Opportunities for clarification of the methodology
- Opportunities for further actions and applications

Developing the TQOL Index

Following the collection of comparative data between countries and the roundtable TQOL meeting in Budapest, the results were used to create a Tourism-Specific Quality of Life Index. A first version of this index was developed following the Hungarian pilot study in 2007 and then the assumed correlations were tested in the other 10 countries. It was subsequently improved after discussions about the international data analysis.

QOL indices can be either theme or country specific. The researchers identified and analysed the experiences of some theme-specific indices, such as Index
of Sustainable Economic Welfare, Measured Economic Welfare, Genuine Progress
Indicator, Human Development Index, Weighted Index of Social Progress and
Happy Planet Index.

The preliminary step in the process of devising a Tourism-Specific Quality-
of-Life Index was the presumption that the analysis of the relationship between
subjective and objective data is possible after normalisation thereof. Objective and
subjective data had to be combined, and a high number of factors needed to be taken
into account. However, access to objective data was often rather limited at (local)
governmental levels, and several factors pertaining to tourists and local residents
were irrelevant to one or the other group. The Index was designed to be relatively
easy to use, reproducible in other contexts, applicable to a wide range of locations,
and useful for planning and management purposes.

The statistical analyses of the questionnaire data were performed using various
(stochastic) statistical methods; the mean of means was calculated separately for
the five main factors (mean of factors), and interdependencies between objective and
subjective data (e.g. by using regressions and correlations) were compared. Based on
the findings of thorough regression analyses it was found that the interrelationships
between the following answers were strong:

- Objective parameters:

  - Number of guest nights per resident
  - Number of visitors per resident

- Subjective parameters:

  - Level of happiness in general
  - Relationship between level of happiness and travel
  - The presence of tourists to the well-being of an individual

Statistically, the five TQOL domains proved to have an impact on the general
QOL domains and parameters. The regression figures for all five were higher than
0.7, i.e. they were significant. The statistical modelling resulted in the following
formula:

$$TQOL - 1 \text{ (tourists)} = -0.04\,V^2 + 1.58\,V - 0.03L^2 + 0.45L + 1.35\overline{ATA} + 1.5\overline{MV}$$
$$+ 1.5\overline{QT} + 1.19\overline{CD} + 1.07\overline{IT} + 8.51$$

$$TQOL - 1 \text{ (local residents)} = -0.06\,V^2 + 0.019\,V - 0.12L^2 + 5.29L + 1.39\overline{ATA}$$
$$+ 1.76\overline{MV} + 1.77\overline{CD} + 1.58\overline{IT} - 7.46$$

where V is the number of guest nights per resident and L is the number of visitors
per resident.

Survey data indicated that more than two of the objective factors should be taken into account when devising the index; however, the reality characterising the data collection methodology of each country does not allow for this. Consequently, only guest nights per visitor and the number of visitors proved to be accessible among all partners. When analysing certain dimensions constituting the index, it may be concluded that:

- data correlate strongly in the case of both segments (i.e. visitors and residents), and
- tourist (T) and visitor (V) data seldom show significant divergence, but this is striking in the case of the ATA and IT dimensions – which, indeed, draws attention to the importance of subjective elements.

The purpose of the development of the Budapest model and the TQOL-Index was to give planners, policy makers and destination managers a tool. This tool aims to provide help identifying the complex nature of tourism and its relationships to QOL. From the detailed data analysis the research team could highlight some major issues any planner or policy maker or destination manager should take very seriously. For example, the data on Harkstede (The Netherlands) stand out from the 12 other locations: both residents and tourists believed that travel essentially decreases their quality of life. The analysis also indicated that visitors to Eger, Siófok (Hungary) and Sunderland (UK) were less happy than permanent residents of the same three towns. The model could identify the major clues or areas that could cause the sometimes surprising findings. Those who use the data, however, should have a more detailed look at the collected information in order to find out what the real drivers can be. The model can provide support for more detailed data analysis, but the index 'only' sends alarming signals about changes and direction of those changes. Further analysis of the available database could provide answers for the unexpected findings in several destinations. The comparison, certainly, has its limitations and the model can be best used on a case-by-case basis.

The TQOL Index results are summarised in the following two charts. Figure 9.7 represents the index for local citizens while Fig. 9.8 represents the related results of visitors.

Noting that while a significant amount of information was collected from every location, the index includes only some of that. The remaining information should also be taken into consideration when one compares the results of the 13 locations, or the data from local citizens and visitors. This happened, for example, when the research partners analysed their respective data. We can see that visitors, not surprisingly, tend to have higher results than local citizens, except in some locations, where data indicate that the existence of tourism improves the QOL of citizens more than that of visitors. This could make planners or managers think that visitors are missing something from their visits.

Without the location- and respondent-specific information, the interpretation of the data could be oversimplified. However, as with every index, the data tabulation should have a certain degree of simplification (e.g. see the World Economic Forum's Tourism Competitiveness Index).

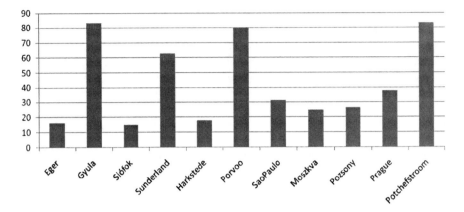

Fig. 9.7 TQOL index results for locals

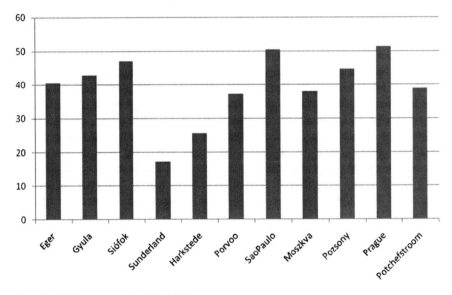

Fig. 9.8 TQOL index results for visitors

Summary and Reflections

We can conclude that the role of travelling and tourism is often undervalued and underrated in general QOL research; therefore, this research-based model represents a significant input into academic, policy and business development.

Based on the value of the calculated index (which may fall between 0 and 100), we may claim that in the case of towns playing a major role in terms of tourism, the quality of life of residents is influenced by tourism to a significant extent, and this impact is altogether positive. However, this relationship is not fixed in stone: different destinations experience different types of relationships, and certain types of tourism, or tourism planned and managed in certain ways could actually decrease QOL.

Despite the overall usefulness of the model for tourism destinations, it may be necessary to refine the model based on some of the comments and criticisms offered by members of the roundtable TQOL group who met in Budapest in 2008, or further testing of the model at destination level. The next step of the refinement of the Budapest model will consider recommendations formulated by the participants:

- Some factors of the five domains should probably be discussed in more detail, since the socio-economic and cultural background of a destination, its residents and the tourists visiting can make a difference to the reading of the results.
- The methodology should probably consider some further, non-tourism-specific, but general QOL factors, e.g. life expectancy, especially if there is no other QOL study available for a destination.
- The findings confirmed that when the general public refers to material goods they think of standard of living or welfare, but when immaterial goods are mentioned they think of happiness or QOL. These differences should be considered in the methodology.
- It may be difficult to compare the responses of visitors to a destination who stay for a relatively short time and will usually feel 'happier' on holiday than at home, and those of permanent residents in a destination. This is especially difficult in cases where the socio-economic backgrounds of residents and visitors are very different (e.g. in developing countries where the residents may rarely, if ever, travel outside the destination).
- Since tourism can be very seasonal in many destinations, one-off data collection may result in considerable bias. Longitudinal data collection is therefore the only way of securing reliable data for any index.
- The number of questions and items in the questionnaire can probably be decreased. However, in social sciences research, especially in QOL studies, questionnaires aiming at the subjective elements of QOL tend to be rather long.

Nevertheless, it is believed that this model goes further than any previous QOL models in measuring the relationship between QOL and tourism, and it can be used as it stands by regional and local government agencies, policy makers and planners. Further refinements can be added by a destination development agency if the results prove to be particularly extreme or alarming (e.g. as in the case of Harkstede in the international pilot test of the Budapest model). This could include further analysis of the subjective questionnaire data, for example. Finally, to reiterate earlier recommendations, the model should be used and analysed on a case-by-case basis and ideally used to collect longitudinal rather than one-off data.

Overall, the importance and timeliness of the topic of the impact of tourism on quality of life is also recognised by international organisations. TQOL and the Budapest model were presented at and were discussed by the OECD Tourism Committee meeting in April 2009 and further discussions and possible actions are to be expected in the near future. Furthermore, the Ministry of Local Government in Hungary is launching a TQOL Blog. They will make the Budapest model data and findings available for members, who will be invited to perfect the methodology further.

Acknowledgement The authors would like to thank the Hungarian Ministry of Local Government and Regional Development for their permission to publish the data from this research which they funded.

References

Ajzen, I., & Fishbein, M. (1977). Attitude-behavior relations: A theoretical analysis and review of empirical research. *Psychological Bulletin, 84*, 888–918.

Andrew, F. M., & Withey, S. B. (1976). *Social indicator of well-being. Americans' perception of quality of life*. New York: Plenum Press.

Ap, J. (1990). Residents' perceptions research on the social impacts of tourism. *Annals of Tourism Research, 17*(4), 610–616.

Campbell, A., Converse, P. E., & Rodgers, W. L. (1976). *The quality of American life*. New York: Russel Sage Foundation.

Cicerchia, A. (1996). Indicators for the measurement of the quality of urban life: What is the appropriate territorial dimension? *Social Indicators Research, 39*, 321–358.

Crotts, J. C., & Holland, S. M. (1993). Objective indicators of the impact of rural tourism development in the state of Florida. *Journal of Sustainable Tourism, 1*(2), 112–120.

Cummins, R. A., et al. (1994). The comprehensive quality of life scale (ComQol): Instrument development and psychometric evaluation on college staff and students. *Educational and Psychological Measurement, 54*(2), 372–382.

Cummins, R. A. (1996). Assessing quality of life. In R. I. Brown (Ed.), *Quality of life for handicapped people* (pp. 116–150). London: Chapman & Hall.

Cummins, R. A. (1997). *The comprehensive quality of life scale – Adult (ComQol-A5)* ((5th ed.)). Deakin University, Melbourne: School of Psychology.

Diener, E., Suh, E., Lucas, R. E., & Smith, H. L. (1999). Subjective well-being: Three decades of progress. *Psychological Bulletin, 125*, 276–302.

Erikson, R. (1993). Descriptions of inequality: The Swedish approach to welfare research. In M. Nussbaum & A. Sen (Eds.), *The quality of life* (pp. 67–87). Oxford: Clarendon Press.

Estes, C., & Henderson, K. (2003). Enjoyment and the Good Life. The less-advertised benefits of parks and recreation. *Park & Recreation, Research Update*. February

Fekete, Z. (2006). Életminőség-koncepciók, definíciók, kutatási irányok. In A. Utasi (Ed.), *A szubjektív életminőség forrásai: Biztonság és kapcsolatok* (pp. 277–301). Budapest: MTA Politikai Tudományok Intézet.

Flanagan, J. C. (1978). A research approach to improving our quality of life. *American Psychologist, 33*, 138–147.

Gnoth, J. (1997). Tourism motivation and expectation formation. *Annals of Tourism Research, 24*(2), 283–304.

Iso-Ahola, S. E. (1982). Toward a social psychological theory of tourism motivation: A rejoinder. *Annals of Tourism Research, 9*, 256–262.

Jeffres, L. W., & Dobos, J. (2003). Perceptions of leisure opportunities and the quality of life in metropolitan area. *Journal of Leisure Research, 25*(2), 203–217.

Johan, N. (2004). Development of a Holistic Tourism and Quality of Life (TQOL) Index. Capturing Residents' and Travellers' Perspectives. *Major Paper Submitted to The School of Hospitality and Tourism Management*. Guelph, Canada.

Kim, K. (2002). The Effects of Tourism Impacts Upon quality of Life of Residents in the Community. *Doctor of Philosophy Dissertation*. Blacksburg, VA: Virginia Polytechnic Institute and State University.

Kovács, Z. (2006). *Dunaújváros és térsége 6: Az életminőség területi különbségei a lakáspiaci dinamizmus tükrében Dunaújváros térségében*. Dunaújváros: Dunaújvárosi Főiskola Térségfejlesztési Kutatócsoport.

Krupinski, J. (1980). Health and quality of life. *Social Science and Medicine, 14A*, 203–211.

Lankford, S. V., & Howard, D. R. (1994). Developing a tourism impact attitude scale. *Annals of Tourism Research, 21*(1), 121–139.

Liu, J., & Var, T. (1986). Resident attitudes toward tourism impacts in Hawaii. *Annals of Tourism Research, 13*(2), 193–214.

Meeberg, G. A. (1993). Quality of life: A concept analysis. *Journal of Advanced Nursing, 18*(1), 32–38.

Neal, D. J., Sirgy, M. J., & Uysal, M. (1999). The role of satisfaction with leisure travel/tourism services and experience in satisfaction with leisure life and overall life. *Journal of Business Research, 44*, 153–163.

Neal, D. J., Uysal, M., & Sirgy, M. J. (1995). Developing a macro measure of quality of life/leisure satisfaction with travel/tourism services: Stage one (conceptualization). In J. Lee et al. (Eds.), *Developments in quality of life studies in marketing* (5, pp. 145–151). DeKalb, IL: Academy Science.

Olfert, S. (2003). *Quality of life leisure indicators*. Saskatoon, SK: CUISR, University of Saskatchewan.

Perdue, R. R., Long, P. T., & Gustke, L. D. (1991). The effect of tourism development on objective indicators of local quality of life. Tourism: Building credibility for a credible industry: Proceedings of the 22nd Annual TTRA Conference. Travel and Tourism Research Association, Salt Lake City, UT., pp. 191–201.

Perdue, R. R., Long, P. T., & Kang, Y. S. (1999). Boomtown tourism and resident quality of life: The marketing of gaming to host community residents. *Journal of Business Research, 44*, 165–177.

Rahman, T., Mittelhammer, R. C., & Wandschneider, P. (2005). *Measuring the Quality of Life across Countries. A Sensitivity analysis of Well-being Indices*. Research paper No. 2005/06 in World Institute for Development Economics Research (WIDER) established by United Nations University (UNU).

Rátz, T., & Puczkó, L. (2002). *The impacts of tourism: An introduction*. Hameenlinna: Hame Polytechnic.

Richards, G. (1999). Vacations and the quality of life: Patterns and structures. *Journal of Business Research, 44*, 189–198.

Sirgy, M. J. (2001). *Handbook of quality of life research: An ethical marketing perspective*. Dordrecht: Kluwer.

Sirgy, M. J., & Cornwell, T. (2001). Further validation of the Sirgy et al.'s measure of community quality of life. *Social Indicators Research, 56*(2), 125–143.

Sirgy, M. J. (2008). Ethics and public policy implications of consumer well-being (CWB) research. *Journal of Public Policy and Marketing, 27*(2), 207–212.

Sirgy, M. J., & Su, C. (2000). Destination image, self-congruity, and travel behaviour: Toward an integrative model. *Journal of Travel Research, 38*, 340–352.

Swan, E., & Combs, L. J. (1976, April). Product performance and consumer satisfaction: A new concept. *Journal of Marketing, 40*, 25–33.

Uysal, M., & Noe, F. (2003). Satisfaction in outdoor recreation and tourism settings, Case studies. In E. Laws (Ed.), *Tourism marketing* (pp. 144–158). London: Continuum Publisher.

Chapter 10
Stakeholder Involvement in Destination Level Sustainable Tourism Indicator Development: The Case of a Southwestern U.S. Mining Town

Donna Myers, Megha Budruk and Kathleen L. Andereck

Introduction

Sustainability interweaves social, cultural, economic, and environmental realms in order to maintain a place, system, or development over time (World Commission on Environment and Development, 1987). With tourism being among the world's largest industries (World Travel & Tourism Council, 2004), scholars and practitioners rely on the concept of sustainable tourism to maximize the positive and minimize the negative impacts to the destination's environmental, economic, social, and cultural resources. Accordingly, sustainable tourism development ensures that the economic viability, ecological integrity, and cultural authenticity of a destination are preserved, thus meeting the needs of present tourists and host regions while protecting and enhancing future opportunities (World Tourism Organization, 1993).

Although many studies conceptualize sustainable tourism, very few incorporate social input from destination communities (Weaver & Lawton, 1999). Local stakeholders are generally the ones who are impacted the most by tourism. Including their perspectives in tourism planning and management is therefore important. One approach for inclusion is through successful stakeholder involvement in the development of tourism-related measurement or indicator systems (Choi & Sirakaya, 2006; Hardy & Beeton, 2001; Timur, 2003). Indicators offer an evaluation output that gives the state or condition of something (Swain & Hollar, 2003). For example, the number of tourism-related jobs provides insight regarding the economic impact of tourism. Such indicators may be used to monitor tourism-related changes over time. Since indicators integrate stakeholder concerns and needs, they provide an avenue for stakeholders to have a voice in tourism development. This holistic and proactive approach to tourism planning and development usually results in increased local ownership and pride, which in turn supports a better tourism product.

D. Myers (✉)
Arizona State University, Phoenix, AZ, USA
e-mail: Donna.M.Myers@asu.edu

M. Budruk, R. Phillips (eds.), *Quality-of-Life Community Indicators for Parks, Recreation and Tourism Management*, Social Indicators Research Series 43, DOI 10.1007/978-90-481-9861-0_10, © Springer Science+Business Media B.V. 2011

This chapter presents a case study that utilizes an interpretive approach to understand the social, cultural, economic, and environmental concerns regarding tourism from a tourism-related stakeholder perspective. Specifically, it identifies destination-level social, cultural, economic, and environmental sustainable tourism indicators from the perspective of residents and business owners associated with the tourism industry in a Southwestern U.S. mining town.

Tourism Impacts

Tourism is a globally occurring activity that intertwines economic, environmental, social, and cultural elements (Timur, 2003). Tourism development occurs at national, regional, and local levels. Within each level, tourism contributes to a variety of economic, environmental, and sociocultural benefits for residents. Examples include income generation, increased currency flow, increased environmental preservation and conservation support, and local involvement in management and decision making (Sirakaya, Jamal, & Choi, 2001). These benefits also extend to tourists who seek destinations reliant on unique resources like historic or archaeological sites, community festivals or rituals, and traditional handmade crafts (Besculides, Lee, & McCormick, 2002; Vander Stoep, 1996). What often results from such tourism is increased community pride and stronger ethnic identity, as well as tolerance for outsiders, thus enhancing understanding among locals, tourism providers, and tourists (Driver, Brown, & Peterson, 1991).

Despite these benefits, in reality, tourism potentially threatens natural and built environments as well as degrades local social systems and culturally authentic lifestyles (Choi & Sirakaya, 2006; Twining-Ward, 2007). Infrastructure damage, strained transportation systems, vandalism, crime, drugs, degradation of a destination's local customs and language, conflict, and fear of foreigners have all been documented (Besculides et al., 2002; Choi & Sirakaya, 2006; Murphy, 1985). These problems often result in a reduced quality of life for local residents (Haralambopoulos & Pizam, 1996) and ultimately negatively affect tourism. While both positive and negative impacts are inevitable, successful sustainable tourism management maximizes the positive impacts and minimizes the negative impacts to the destinations' environmental, economic, social, and cultural resources (Lane, 1994).

Stakeholder Theory in Sustainable Tourism

Stakeholder involvement is an integral element in sustainable tourism development (Walsh, Jamrozy, & Burr, 2001). In the sustainable tourism context, local businesses, chambers of commerce, convention and visitors bureaus, governments, community organizations, local residents, and tourists are considered stakeholders (Sautter & Leisen, 1999; Timur, 2003). Parallel to business management, stakeholder theory could be a framework within which sustainable tourism exists (Robson & Robson,

1996). Specifically, stakeholder theory may be the basis for incorporating stakeholder concerns in tourism development. This is achieved when decision makers understand that stakeholders too hold a legitimate interest in tourism development, and their opinions have intrinsic value, regardless of the nature of their values or beliefs (Robson & Robson, 1996; Sautter & Leisen, 1999). Besides acknowledging stakeholders as legitimate partners, stakeholder theory requires that all stakeholders receive equal and simultaneous attention in the policy- and decision-making process (Hardy & Beeton, 2001). Thus, stakeholders, rather than being considered as intermediaries between a destination and its visitors, play an active role in tourism planning and development (Sautter & Leisen, 1999).

Stakeholder theory applications in sustainable tourism development are few (e.g., Hardy & Beeton, 2001; Timur & Getz, 2008; Timur, 2003), but have increased understanding regarding stakeholder perceptions of the tourism product, impacts, planning (present and future), and marketing (Hardy & Beeton, 2001). Beyond perceptions, stakeholder theory has proved useful in exploring barriers to successful collaboration and communication among various groups (Timur, 2003) and in analyzing stakeholder characteristics, values, and relationships (Timur & Getz, 2008).

Collectively, these studies provide some understanding of stakeholder perceptions, concerns, and relationships. However, successful implementation of sustainable tourism requires that stakeholder concerns are not only understood but also incorporated into management policies. One approach to this is through the development of a stakeholder-based measuring and managing system (or indicator system), which may then be used to monitor tourism-related impacts (Go, Milne, & Whittles, 1992).

Sustainable Tourism Indicators

Indicators are defined as measurable, manageable variables that characterize the quality of a variable of interest (Manning, 1999; Swain & Hollar, 2003). For instance, the number of tourism-related jobs is a measurable and manageable variable that may be used to trace the economic impact of tourism. Several criteria for good indicators have emerged, including that it has policy implications and is specific (i.e., describes a set of currently existing situations and conditions), reliable, valid, manageable, as well as efficient and effective to measure (Manning, 1999).

Sustainable tourism indicators exist at the national, regional, and destination level (World Tourism Organization, 2004). However, destination level indicators that focus on social, cultural, economic, and environmental concerns are especially useful for two reasons. First, although national level indicators such as Gross Domestic Product provide a framework for regional- and destination level indicators, often national indicators are too broad to be used by stakeholders who are concerned with specific attributes of their region or destination. Since destination level indicators consider the unique character of a destination, they are deemed the most integral to community stakeholders (Sirakaya et al., 2001). Second, while indicators have been useful in understanding and responding to tourism-related impacts, these indicators (e.g., income generation) have generally focused on economic impacts. Sustainable tourism indicators focus on all dimensions of sustainability, especially social and cultural, thus providing stakeholders with a more holistic understanding of their destination's unique tourism development potential and associated impacts (Twining-Ward, 2007).

Thus, stakeholder involvement is integral in sustainable tourism development; however, few studies have applied stakeholder theory in sustainable tourism and much remains to be explored, especially with respect to incorporating stakeholder concerns into tourism management and policy. One approach is to develop stakeholder-based tourism indicators which may then be used to track tourism-related impacts. Tourism scholars suggest developing indicators that not only focus on the economic dimension but also include environmental, social, and cultural dimensions.

Sustainable Tourism Indicator Development

Mammoth, Arizona

Mammoth, Arizona, was founded in 1876 and incorporated as a town in 1958. Located about 100 miles southeast of Phoenix, Arizona, the town sits at the base of the Galiuro Mountains, among Sonoran Desert landscape that includes washes, hills, and canyons. Mammoth's history is closely linked to copper ore mining in the area. Several homes and artifacts in Mammoth were originally located in a mining town called Tiger. The former township of Tiger sits on a hill that overlooks Mammoth. Tiger was also home to a mining operation that holds the historical relics and stories significant to Mammoth's history. Today, Tiger no longer exists.

Research Approach

Data were gathered through semistructured open-ended interviews with local business owners and residents during January 22–26, 2009. The researcher utilized an interpretive approach which is useful when examining research questions that are multidimensional, exploratory, and incorporate qualitative data collection techniques such as interviews and collaborative dialogue (White, 2007). Sampling followed a theory-based approach, whereby participants are selected based on their ability to help develop an emerging theory (Miles & Huberman, 1994). The sample was based on the individual respondent's status as a local stakeholder in Mammoth's tourism development. Specifically, business owners and local residents who either affect or will be affected by tourism development were considered. An additional consideration was their ability to offer insight into both the social systems and the cultural customs of the community (Glaser et al., 1968). Recruiting stakeholders involved both purposive and snowball sampling (Henderson, 2006). Purposive sampling involves arbitrarily selecting study participants based on evidence that the particular sample represents the total population (Henderson & Bialeschki, 2002). Snowball sampling involves asking participants to recommend other potential participants. These sampling techniques ensured representativeness among business owners and residents. Sampling occurred until data saturation was achieved, i.e., when data became repetitive and no new or relevant information emerged (Henderson & Bialeschki, 2002).

Interview Protocol

Eighteen interviewees were approached. The open-ended semistructured interviews lasted up to 25 min. Questions focused on stakeholder perceptions of tourism-related impacts. Additionally, participants were asked about salient tourism-related impacts that could be tracked throughout the tourism development phases. Finally, participants were asked about various tourist attractions in Mammoth. The interviews were digitally audio recorded.

Analysis

The recordings were transcribed shortly after each interview, and relevant portions of the transcripts were coded. Coding is "reducing the words to numbers, short phrases or to short descriptions" primarily relying on descriptive words codes (Henderson & Bialeschki, 2002 p. 304). Specifically, the researcher developed a list of start codes or explanatory groupings of anticipated responses to the interview questions (for a detailed description of the coding process, see Miles & Huberman, 1994). Once completed, a second researcher applied the start codes to four of the 15 transcripts. The two researchers then discussed any differences in interpretation. Based on this, new coding categories or descriptive codes were created. This process guaranteed intra-rater and inter-rater reliability (Miles & Huberman, 1994). The descriptive codes were used to explore emergent themes which are described next.

Sample Description and Themes

Sample characteristics (including demographics) and descriptive coding results (themes) related to the social, cultural, economic, and environmental dimensions of tourism-related changes are presented below. These themes were used to identify destination-specific sustainable tourism indicators.

Although tourism literature indicates several destination level stakeholder groups, this study incorporated only two groups: business owners and residents. Fifteen stakeholders agreed to be interviewed. Of these, nine were business owners (five of whom were female). Of the remaining six residents, three were female.

Descriptive coding resulted in four dimensions of sustainable tourism indicators: social, cultural, economic, and environmental. The dimensions and their respective

Table 10.1 Themes and associated indicators for the social, cultural, economic, and environmental dimensions of sustainable tourism indicators

Dimension	Themes	Indicator
Social	Town appearance and cleanliness (6[a])	# of businesses with clean appearance
		# of beautification and city improvement projects
	Stakeholder involvement (4)	Active participation in community events
		Attendance at town council and planning and zoning meetings
		Youth participation in community events and programs
		Collaboration among tourism development partners
	Quality of life (9)	# of affordable homes
		# of new residents
		# of sports events and recreation programs
		# of local gathering places and events
Cultural	Local pride (5)	Existence of local pride
		Town cleanliness
		Community collaboration
		# of new residents
		Active participation in events
		# of jobs
		Increase in local commerce
	Local identity (7)	# of restored buildings
		Presence of mining history museum
		Presence of photo archives
		Presence of transcripts and oral history archives
		Ore cart memorial
Economic	Increase in local commerce (13)	# of businesses
		# of jobs
		# of franchise stores
		# of business license restrictions
	Local spending (1)	Sales receipts
Environmental	Parks (2)	# of parks
		# of improvement projects at parks
	Abandoned building (2)	# of abandoned buildings

[a]Numbers in parenthesis indicate number of times the theme was referenced in the interviews.

themes are listed in descending order of frequency of total appearances in the interviews: social (49), economic (45), cultural (24), and environmental (6). The following are exemplary quotes from the most frequently occurring themes of each dimension. Indicators were developed from these themes. A summary of these themes and associated indicators is presented in Table 10.1.

Social Dimension Themes and Indicators

The descriptive coding resulted in three themes related to social changes. These themes are highlighted below with illustrative quotes.

Town cleanliness and appearance: Town cleanliness and general physical appearance, especially along Mammoth's main street, State Route 77, were referenced seven times throughout the interviews. Business owners were particularly vocal on this matter as it relates to Mammoth's ability to attract tourists: "I think the main issue with the whole tourism thing is that if the town is not clean, tourists don't come. It has to be presentable, people just drive through" stated one business owner.

Business owners' suggestions on how to achieve cleanliness and improve the town's general appearance primarily applied to other businesses and not specifically their own. For example, one business owner stated, "I think they (other business owners) could clean up their businesses, paint them and make them presentable, there are a lot (of businesses) in town that are not presentable."

In addition to individual business cleanliness, the same business owner shared her thoughts about the main street, where most of Mammoth's businesses are situated:

> There's a lot of stuff along the main drag that I can see the need for changes to. Not to mention off the main drag, what a bunch of trash heaps there are in this town. It just needs to be cleaned up, I don't look for tourism to even come to this area because of that.

In addition to individual businesses' cleaning and trash removal efforts, a business owner and a resident suggested beautification and city improvement projects. "If they kept the town, plants some stuff on the side of the road, some trees, some flowers," stated a business owner. The resident suggested more parks and commended the new streetlights and sidewalk on the main street. Also, business owners and residents considered the recently erected Miner's Memorial along the State Route 77, an excellent improvement to Mammoth's main street. Both residents and business owners shared an interest in improving the town's appearance for tourists.

Stakeholder involvement: Stakeholder involvement, a second theme, emerged four times in the interviews. A business owner shared a couple of suggestions regarding ongoing stakeholder involvement in Mammoth: "So more active participation in community events because they don't even come to the Planning and Zoning or Council meetings, which are important to their community." When asked to further explain stakeholder involvement in Town Council and or Planning and Zoning meetings, she offered the following explanation:

> I feel that the Planning and Zoning is the front line for the town. More so than Council. Any changes that happen in the town, especially if they're significant have to go through Planning and Zoning first. More participation (in both is needed).

This same business owner suggested that collaboration among all stakeholders was also necessary. Finally, another suggestion related to youth participation in community events and programs.

Quality of life: The last theme related to the social dimension, quality of life for Mammoth residents, appeared nine times. Stakeholders spoke about safety, lack of crime, good schools and jobs, peacefulness and tranquility, camaraderie, as well as the breathtaking scenery as reasons they liked living in Mammoth. When asked about what tourism-related impacts to quality of life it was important for the town to keep track of, stakeholders mentioned the number of affordable homes and new residents. A business owner and two residents shared the importance of affordable housing if tourism were to develop in Mammoth:

> That the prices of homes, they're going to try to make it to where the people who live in this town have first choice at the brand new homes. That is what we are waiting on. And if the economy got better, my wife and me were talking about buying a brand new home here.

Two respondents stated that it is important to track new homes and residents that come into town as a result of tourism development: "Besides bringing more people in, they might like the situation and the area to move in to get more neighbors, make more friends. Definitely, you know with people stopping through. People could buy a place down here."

Besides affordable homes and new residents, sporting events and recreation opportunities were another quality of life indicator revealed through descriptive coding. A resident and a business owner suggested that Mammoth should monitor city/regional sport events and recreation programs numbers as a possible tourism-related impact. Stakeholders felt the current sports events and recreation programs could be expanded as tourism develops. For instance, two business owners suggested promoting Mammoth's existing motocross facility to tourists. Another mentioned soccer events at the local park that draw regional visitors. Additionally, the same business owner discussed some of the parks improvements that could be made:

> Well okay, we have several parks, ball parks, like the Little League here and we've got a soccer field down at the very end, we don't have any bathrooms down there, we don't have an entrance off the highway. We draw a lot of people when we have those soccer games down there.

The number of local gathering places and events was the final quality of life indicator identified through descriptive coding. Stakeholders noted a lack of gathering places and events in Mammoth conveyed a diminished quality of life. The responses by two business owners to the question "what social attractions do you think Mammoth currently offers to tourists" illustrate this point: "That probably is the weakest point here, there's really not too much. You know we've got the restaurants and everything and that's about it. There are no more fiestas that go on, there's not really any good hang outs, and I doubt that there is a social thing in Mammoth that could attract them."

Despite the current situation regarding social attractions, business owners and residents stated that the number of both local gathering places and events could improve the quality of life in Mammoth and are important to measure as tourism develops. One business owner said, "We could bring more people in town with more of that type of business (a social gathering place like a bar)." Another resident spoke about both gathering places and events:

> There need to be more festivities/celebrations of some sort to make the town better. They do need a bar, to blow off steam, even a dance hall. People work hard either in Tucson or the smelter or the mines. They want to go to the bar, watch the game, watch the fight. There might be a dance or a concert of some sort.

Cultural Dimension Themes and Indicators

Two themes emerged under the cultural dimension of tourism-related impacts.

Existence of local pride: The existence of local pride is a unique cultural indicator in that there are no clear objective measures. The following quotes reveal stakeholders' suggestions on measuring pride related to the social and economic indicators. These emerged five times in the interviews.

One business owner connected town cleanliness and pride:

> "Our community should be getting looking better. Hopefully would give people a little bit more encouragement in our self-pride. It would also kind of motivate me more."

She discussed her experience in the context of other Mammoth stakeholders. By doing this, she implied that community collaboration to improve the economy, would contribute to local pride:

> "Jobs, again, income to the community. Perhaps people who have not been able to do the things that they want to, because of lack of money. That perhaps helps that way. Maybe just pick up the spirits to where they wanted to work together as a community to make it work."

Another resident expanded upon the connection between vibrant economic growth and local pride by stating "For it to grow, to expand and to be able to say 'hey, I'm proud of the town of Mammoth.' Because it got better, the economy got better, more tourism came in. More people want to move in, more people want to build motel rooms. More people want to stay here."

In addition to economic growth, a business owner discussed local involvement as a source of pride:

> "So more active participation in community events and its getting people encouraged and involved."

Another business owner discussed the importance of not only having involved current residents, but also having new members of the community contribute. This business owner touched on the importance of new ideas, energy, and collective experience in order to infuse pride into a town that currently lacks vibrancy. A related theme to pride was local identity.

Existence of local identity: Stakeholders were asked about Mammoth's existing tourist attractions and their importance. Most were able to point out existing attractions that the town possesses and articulated the sense of identity in seven excerpts. Some highlights of the attractions included the Galiuro Mountains, the Mexican culture, and the town's history. "The Galiuro Mountains are just so beautiful. That was one of the final selling points on our home for us was," "The mariachi bands and the dancers, that's a tradition that has stuck," "The old courthouse. I think the owner still has a lot of the old transcripts from the court cases that occurred there. It would be awesome, if they could redo the old courthouse."

In order to preserve this cultural identity, a business owner suggested that buildings in town be restored: "I'd like to see all the old buildings refurbished. Brought back to life again, there's one downtown that's been abandoned. It's in an old adobe building, I'm sure the structure is not that great, but if they would go in and clean it up and make it look presentable and safe."

In addition, several stakeholders suggested that mining relics and town history be archived and displayed in a museum. Others suggested that there should be photo archives in addition to remaining transcripts and oral accounts. One resident talked excitedly about oral histories:

> "This town is full of history, case in point, this guy right here, he'll tell you everything in town that happened. Along main street, the bars. He'll remember back in the day this department store used to be open, this dance hall used to be open."

A couple of stakeholders acknowledged that local Ore Cart Memorial that honors miners and their families is a positive way to display Mammoth's identity.

These stakeholders' suggestions serve as the basis for the indicators created regarding cultural pride and identity. The results reveal that while difficult to directly measure, the two cultural themes were linked to the social and economic dimensions.

Economic Dimension Themes and Indicators

Three themes emerged under the economic dimension of tourism-related impacts.

Increase in local commerce: This was among the most frequently occurring theme with 13 references. A business owner pointed out the effects of the current lack of local commerce by saying, "Perhaps people who have not been able to do the things that they want to, because of lack of money." Another offered a simple response about keeping track of local commerce in Mammoth, "More people, more jobs, a little bit more economy." A different business owner, when referring to cleaning up the town and making businesses more presentable, pointed out that an increase in the local economy would help, but local stakeholders have work to do before tourism would be successful in Mammoth.

Based on this theme, the most frequently occurring economic indicator related to the number of businesses. Five business owners and three residents spoke about the importance of tracking the number of businesses during tourism development. A resident responded by saying "People come, so keep track of how business goes up." A business owner mentioned tracking specific types of businesses, "It would be great to have more restaurants, but there really is not much in town, we have 2 or 3 restaurants in town right now."

Mammoth stakeholders identified another important economic indicator: number of jobs. "More tourism creates more jobs, more jobs creates more for the people here who desperately need it." Furthermore, this same resident said:

> But if they brought in tourism and opened up businesses involved in tourism, it would maybe open up more jobs for people. That would need part time or full time employment instead of having to depend on going out or on the government, as far as welfare, unemployment, and social security.

Stakeholders also suggested that the number of franchise stores in Mammoth as an important economic indicator. Currently, there are no major chain stores in Mammoth. A resident summed this up by saying:

> I think they need probably, a bookstore, or maybe a movie theater. More shopping. As far as places to get gas there is only two places in town. When one of them shuts down, the other one is going to go. There's no supermarket at all.

A final economic indicator under this theme pertained to people's ability to start a business. Two business owners highlighted their concern about the number and relevance of business license restrictions in Mammoth: "Making it easier on them to open up a business. They have so many restrictions for a small town they don't need. It's closing the doors on businesses from opening up," and "A lot of planning and zoning rules that don't really pertain to a small town like this... they brought in rules from other towns that don't pertain to this and make it harder for anyone to keep running a business around here," revealing the need for tracking in this area.

Local spending: This was the final theme that emerged under the economic dimension of tourism-related impacts. The one passage that captures the nature of this theme is below:

> I'm hoping tourism comes to the Town of Mammoth because they need it big time, because like I said, it would help, create jobs, which the people would need to help support their families maybe reinvest back in the town again, buying more stuff. If not the more and more people are unemployed here they have to go find jobs else and not spend their money in town and spend it elsewhere.

The following and final section presents themes and indicators from sustainable tourism's environmental dimension.

Environmental Dimension Themes and Indicators

Stakeholders identified two themes related to this dimension, parks and abandoned buildings.

Parks: Two stakeholders thought it is important to track the number of parks as tourism develops. In addition to becoming an attraction for visitors, one business owner pointed out that everyone would benefit from the parks: "Well okay, for more attractions, if we could get a grant to finish up the park right below the cemetery, maybe put a bathroom, even just one for everybody. Now like I say, people from Oracle, San Manuel, we get together and everyone comes in with their kids. It would be to everyone's benefit." A resident also suggested keeping track of park improvements.

Abandoned buildings: Abandoned buildings were referenced two times. Views varied from revitalizing to razing the abandoned buildings. Regardless of the outcome, stakeholders agreed that the number of abandoned buildings in Mammoth should be tracked. One business owner suggested revitalization of the abandoned properties, relating back to the suggestion that old restored properties can be used to attract visitors.

Another business owner pointed out the challenges in dealing with abandoned properties, "But if its abandoned properties, you can get people to clean them up, but how do you get them to raze the buildings?" This statement reiterates the need to improve upon some of the visual blights in Mammoth. Specifically, the challenge of abandoned buildings falls into the environmental dimension of sustainable tourism.

Incorporating Indicators into Sustainable Tourism Development and Management

Mammoth serves as an example of a developing destination whose stakeholders seek to achieve economic enhancement through tourism development. Themes from four dimensions of sustainable tourism (social, cultural, economic, and environmental) emerged during interviews with community members. These were used to identify indicators of sustainable tourism that may be incorporated into Mammoth's tourism development plan. Specifically, Mammoth stakeholders may utilize these indicators to gain precise measurements of the state of their economy, quality of life, town cleanliness, local involvement, and cultural preservation efforts. These indicators can serve as a frame of reference for the current and future state of affairs in Mammoth and aid in monitoring tourism-related changes.

Respondents identified stakeholder involvement as one of the three themes within the social dimension. Specifically, stakeholders identified the indicator, involvement in Planning and Zoning or Town Council Meetings and community events. These findings are consistent with studies that present a framework for stakeholder involvement (Choi & Sirakaya, 2006; Twining-Ward, 2007). Additionally, in this study, collaboration among community members was identified as an indicator within the stakeholder involvement theme. This finding is also consistent with the literature (Bramwell & Lane, 2000; Jamal & Getz, 2000) and reiterates Timur and Getz's (2008) finding that destination level sustainable tourism depends upon effective collaboration.

Quality of life emerged as an important theme within the social dimension. Quality of life for residents at tourism destinations was a concept present in early literature on sustainable tourism. For instance, Green and Hunter (1992) and Hunter and Green (1995) made the connection between resident quality of life and visitor experience enhancement. The quality of life indicators within the social dimension identified in this study mainly relate to improvements to the town that could benefit both residents and visitors. For example, two indicators identified were number of local places and events as well as number of sports events and recreation programs. Mammoth stakeholders should consider this dynamic when developing tourism in their town. It is important that stakeholders create or revitalize attractions that will not only successfully attract visitors, but at least maintain or improve their own quality of life. A relevant question that Mammoth stakeholders can ask and determine among themselves is "how will this attraction or development contribute to the local stakeholders' quality of life?" Again, Mammoth stakeholders can reference the indi-cators identified in this study to answer this question and gauge tourism's impact on their quality of life. Once again, it will be important that stakeholders add to and revise the indicators identified in this study as they build attractions and welcome visitors.

Mammoth stakeholders identified two other quality-of-life indicators similar to ones identified by Choi and Sirakaya (2006): number of affordable homes and number of new residents. This suggests that perhaps there are universal indica-tors of sustainable tourism. It is important that sustainable tourism researchers and practitioners not "reinvent the wheel" when identifying relevant indicators for their destination. The consistencies found between indicators in this study and those existing in the literature reveal that there are some indicators, like number of jobs, that are universal. Sustainable tourism professionals, researchers, and stakehold-ers should reference existing indicators before setting out to determine their own unique destination indicators. That is not to say that stakeholders should not go about identifying the most appropriate indicators that incorporate the distinctive attributes of their destination. On the other hand, stakeholders should carefully con-sider the unique resources of their destination in order to develop destination level indicators that will accurately reflect the tourism-related impacts in their area.

Indicators from the cultural dimension identified here are also consistent with the literature. Specifically, Hardy and Beeton (2001) discovered that local pride ensures a better tourism product. They found that this pride could be a result of stakeholder involvement and ownership of the local tourism industry. Mammoth stakeholders equated pride with economic growth through tourism that results in more jobs, income, and local investment. Mammoth stakeholders should continue to incorpo-rate these sources of pride as the research implies that boosting local pride is crucial to building a successful tourism industry. In order to build pride and a successful tourism product at their destination, Mammoth stakeholders should constantly refer and regularly update the indicators associated with existence of local pride. This is a suggestion for all tourism planners and managers, as well.

The indicators identified were based on select local stakeholders' perceptions and are destination specific. Each indicator identified is measurable and manageable, both of which are criteria for good indicators (Manning, 1999). The next stage is to formulate standards based on the indicators identified in this study. Standards of quality are specific goals that Mammoth stakeholders could formulate to determine success or failure within each indicator. For example, Mammoth stakeholders could specify the number of tourism-related jobs that they consider appropriate for their destination.

Conclusion

Stakeholder theory has been proposed as a framework within which sustainable tourism exists (Robson & Robson, 1996). Additionally, stakeholder perspectives regarding tourism-related changes have been identified as a valuable resource in sustainable tourism development (Timur, 2003). Despite this, only a few studies address stakeholders in sustainable tourism research. Even fewer incorporate stakeholder perspectives into sustainable tourism development (Hardy & Beeton, 2001). This case study examined select stakeholder perceptions regarding sustainable tourism in Mammoth, Arizona, using an interpretive method. Through interviews with current business owners and residents in Mammoth, a number of sustainable tourism indicators were identified. Stakeholders in Mammoth may utilize these indicators to track tourism-related impacts in their town over time. These indicators are meant to gauge the current and future state of the environment, economy, and socioeconomic conditions in Mammoth, and therefore may be modified to suit stakeholder needs throughout all stages of tourism development. In addition to use by Mammoth stakeholders, these indicators can serve as a model of destination level indicators for communities involved in sustainable tourism.

References

Besculides, A., Lee, M. E., & McCormick, P. J. (2002). Residents' perceptions of the cultural benefits of tourism. *Annals of Tourism Research, 29*(2), 303–319.

Bramwell, B., & Lane, B. (2000). *Tourism collaboration and partnerships: Politics, practice and sustainability.* Australia: Channel View Books.

Choi, H. C., & Sirakaya, E. (2006). Sustainability indicators for managing community tourism. *Tourism Management, 27*(6), 1274–1289.

Driver, B. L., Brown, P. J., & Peterson, G. L. (1991). *Benefits of leisure.* State College, PA: Venture Pub.

Glaser, B. G., Strauss, A. L., & Strutzel, E. (1968). The discovery of grounded theory: Strategies for qualitative research. *Nursing Research, 17*(4), 364.

Go, F. M., Milne, D., & Whittles, L. J. R. (1992). Communities as destinations: A marketing taxonomy for the effective implementation of the tourism action plan. *Journal of Travel Research, 30*(4), 31.

Green, H., & Hunter, C. (1992). The environmental impact assessment of tourism development. In P. Johnson & B. Thomas (Eds.), *Perspectives on tourism policy* (29–48). London: Mansell.

Haralambopoulos, N., & Pizam, A. (1996). Perceived impacts of tourism the case of Samos. *Annals of Tourism Research, 23*(3), 503–526.

Hardy, A., & Beeton, R. J. S. (2001). Sustainable tourism or maintainable tourism: Managing resources for more than average outcomes. *Journal of Sustainable Tourism, 9*(3), 168–192.

Henderson, K. A. (2006). *Dimensions of choice: A qualitative approach to recreation, parks, and leisure research* (2nd ed.). State College, PA: Venture Publishing, Inc.

Henderson, K. A., & Bialeschki, M. D. (2002). *Evaluating leisure services: Making enlightened decisions.* (2nd ed.). State College, PA: Venture Publishing.

Hunter, C., & Green, H. (1995). *Tourism and the environment: A sustainable relationship?* London, New York: Routledge.

Jamal, T., & Getz, D. (2000). Community roundtables for tourism related conflicts: The dialectics of consensus and process structures. In B. Bramwell, & B. Lane (Eds.), *Tourism collaboration and partnerships: Politics, practice and sustainability.* Clevedon: Channel View Books.

Lane, B. (1994). Sustainable rural tourism strategies: A tool for development and conservation. *Journal of Sustainable Tourism, 2*(1–2), 102–111.

Manning, R. E. (1999). *Studies in outdoor recreation: Search and research for satisfaction* (2nd ed.). Corvallis, OR: Oregon State University Press.

Miles, M. B., & Huberman, A. M. (1994). *Qualitative data analysis: An expanded sourcebook* (2nd ed.). Thousand Oaks, CA: Sage Publications.

Murphy, P. E. (1985). *Tourism: A community approach.* London: Routledge.

Robson, J., & Robson, I. (1996). From shareholders to stakeholders: Critical issues for tourism marketers. *Tourism Management, 17*(7), 533–540.

Sautter, E. T., & Leisen, B. (1999). Managing stakeholders a tourism planning model. *Annals of Tourism Research, 26*(2), 312–328.

Sirakaya, E., Jamal, T. B., & Choi, H. S. (2001). Developing indicators for destination sustainability. In D. Weaver (Ed.), *The encyclopedia of ecotourism* (pp. 411–432). New York: CAB International.

Swain, D., & Hollar, D. (2003). Measuring progress: Community indicators and the quality of life. *International Journal of Public Administration, 26*(7), 789–817.

Timur, S. (2003). Sustainable urban tourism: A stakeholder theory perspective. *Tourism: An International Interdisciplinary Journal, 51*(2), 111–126.

Timur, S., & Getz, D. (2008). A network perspective on managing stakeholders for sustainable urban tourism. *International Journal of Contemporary Hospitality Management, 20,* 445–461.

Twining-Ward, L. (2007). Adapting the indicator approach – practical applications in the south pacific. In R. Black, & A. Crabtree (Eds.), *Quality assurance in ecotourism* (pp. 116–135). Cambridge, MA: CAB International.

Vander Stoep, G. (1996). Perception of status of Michigan as a heritage tourism state: Results of an eleven month telephone survey. *Proceedings of the 1996 Northeastern Recreation Research Symposium. USFS General Technical Report NE-232.* Radnor, PA: USFS.

Walsh, J. A., Jamrozy, U., & Burr, S. W. (2001). Sense of place as a component of sustainable tourism marketing. In S. F. McCool, & R. N. Moisey (Eds.), *Tourism, recreation and sustainability: Linking culture and the environment.* Oxford: CABI Publishing.

Weaver, D. B., & Lawton, L. (1999). *Sustainable tourism: A critical analysis.* Gold Coast: CRC for Sustainable Tourism.

White, D. D. (2007). An interpretive study of Yosemite national park visitors' perspectives toward alternative transportation in Yosemite Valley. *Environmental Management, 39*(1), 50–62.

World Commission on Environment and Development (1987). *Our common future.* Oxford: Oxford University Press.

World Tourism Organization (1993). *Sustainable tourism development: A guide for local planners.* Madrid: World Tourism Organization.

World Tourism Organization (2004). *Indicators of sustainable development for tourism destinations: A guidebook.* Madrid: WTO.

World Travel & Tourism Council (2004). *World travel & tourism: A world of opportunity, the 2004 travel and tourism economic research.* London: WTCC.

Chapter 11
The Trials and Tribulations of Implementing Indicator Models for Sustainable Tourism Management: Lessons from Ireland

Kevin Griffin, Maeve Morrissey, and Sheila Flanagan

Background

This chapter is composed of two parts. Part A outlines the development of the *DIT-ACHIEV Model for the Sustainable Management of Tourism* and part B presents a discussion on its implementation, with particular emphasis on how community actors have been involved in its testing and implementation. This model of sustainable tourism indicators has been developed by the School of Hospitality Management and Tourism, Dublin Institute of Technology (DIT) and is endorsed by the Irish Environmental Protection Agency and Fáilte Ireland (the National Tourism Development Authority of the Republic of Ireland). The model is designed to mitigate the negative impacts of tourism and guide a destination towards a broad range of activities which will encourage movement towards true sustainability. The model comprises six fields of interest, the initials of which lead to its name:

- Administration
- Community
- Heritage
- Infrastructure
- Enterprise
- Visitor

Part A: Development of the Model

Rationale

Sustainable development encompasses not only environmental protection but also economic development and social cohesion (Kelly, Sirr, & Ratcliffe, 2004). Agenda

K. Griffin (✉)
Dublin Institute of Technology, Dublin, Ireland
e-mail: Kevin.griffin@dit.ie

M. Budruk, R. Phillips (eds.), *Quality-of-Life Community Indicators for Parks,
Recreation and Tourism Management*, Social Indicators Research Series 43,
DOI 10.1007/978-90-481-9861-0_11, © Springer Science+Business Media B.V. 2011

21 (arising from the United Nations Conference on Environment and Development (UNCED) in Rio de Janeiro, 1992) represents a blueprint intended to set out an international programme of action for achieving sustainable development for the twenty-first century. Since UNCED, this has been incorporated, at a practical level, into EU and Irish policy formation, with Local Agenda 21 advocating the belief that sustainable development will be most successful if initiated at the local level.

A major challenge to achieving sustainability in the development of Irish tourism was the unprecedented growth in the industry from 1985, which put pressure on physical infrastructure and risked placing severe stress on the quality of the environment – the very basis of the Irish product (Flanagan et al., 2007). Mindful of this, a main challenge for tourism in Ireland is to develop and promote a product that is environmentally, socially and economically responsible and in due course, an active player in the drive for sustainable development. According to Ireland's Environmental Protection Agency (EPA) EPA (2000), this requires existing and new tourism developments to incorporate adequate protection measures to enhance the quality of the existing environment and to mitigate tourism destination impacts. Applying Agenda 21 principles, this must initially be done at the local level.

Recently, a section of *Ireland's Environment 2008* (EPA, 2008a) was dedicated to tourism and travel combined. This, fourth State of the Environment report, calls for tourism impacts to be closely monitored due to direct and indirect impacts on the environment. Building on previous reports (EPA, 2004, etc.) this recognises tourism as an important source of investment and employment in Ireland, particularly in rural regions. These various national reports acknowledge how tourism interacts closely with other policy areas – transport, energy, environment, regional planning, business and trade – and the need therefore to co-ordinate and integrate policies. Additionally, all stakeholders in the tourism sector, at national, regional and local level, have a part to play in working towards sustainability. The challenge, however, is that many of the potential actors have no idea how this can be attained; thus, there is a need for clear and concise direction from the tourism development agencies.

Managing for Sustainable Tourism

The regulation and management of an area is important to mitigate against negative tourism impacts. Williams and Gill (1993) note a need to limit and control tourism, which may threaten sustained use of limited resources. In order to determine where such actions need to be taken, a thorough audit of tourism impacts is required. Glasson, Godfrey, and Goodey (1995: p. 51) in assessing urban tourism impacts advised that the multitude of "'hard' objective or quantifiable dimensions and 'soft' or more subjective, qualitative perspectives" are closely interlinked in a network or 'web' of elements which must be considered in totality before one can understand and manage visitor impacts. Their web or network of elements (see Fig. 11.1) which is split between these quantifiable dimensions and qualitative perspectives was a major inspiration for the development of the DIT-ACHIEV model. In the carrying capacity web, the quantifiable dimensions concern ecological systems, physical

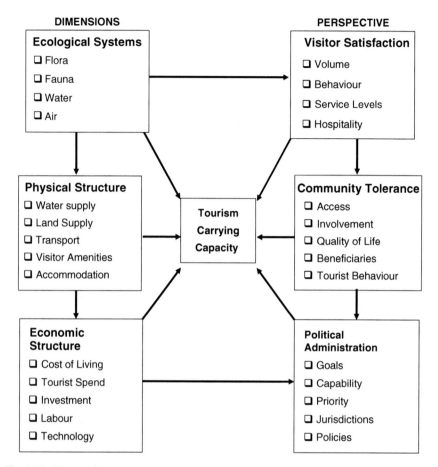

Fig. 11.1 The carrying capacity web (Glasson et al., 1995)

structures and economic structures, and the qualitative perspectives relate to visitor satisfaction, community tolerance and political administration.

A challenge in developing a viable model of indicators is allowing sufficient breadth to generate an all-encompassing understanding of the issues under investigation, while remaining suitably focussed to target specific problems and challenges. A further concern is how to weave less tangible but important issues such as value judgements into the decision-making process of several interconnected tourism organisations. Consequently there is a need to blend the investigation of themes such as community partnerships, cooperatives and corporate social responsibility with more quantitative measurements such as water quality and income generation to achieve a true understanding of sustainable tourism. The original research undertaken in developing the DIT-ACHIEV model addressed these issues and proposed that organisations can respond effectively to changes in tourism, thereby mitigating negative impacts and leading the industry in the direction of sustainability.

Sustainable Tourism Indicators

The UN World Tourism Organisation (2004,: p. 8) defines indicators as

> measures of the existence or severity of current issues, signals of upcoming situations or problems, measures of risk and potential need for action, and means to identify the results of our actions.

While both quantitative and qualitative measurements are required for a full assessment of a destination, care must be taken in both instances to ensure that indicators do not contain value judgements and are developed on a scientific basis. Thus, the ability of all variables to reflect given phenomena must be rigorously established. Engagement with subject experts to undertake extensive validation of appropriate indicators comprised a large share of the work in the initial development of the DIT-ACHIEV model.

Characteristics of Indicators

A properly functioning system of indicators should provide information which helps interested parties to communicate, negotiate or make decisions in a manner which is conducive to the efficient achievement of objectives. In tourism terms, the overall objective will be the sustainable development of a destination. Therefore, in order to be useful, tourism development indicators must fulfil a variety of important functions. The MEANS programme, which reflects the findings of research by the European Commission (1999) into evaluating socio-economic programmes, identifies eight criteria for viable indicators (see Table 11.1).

These guidelines provided useful clarity in refining the DIT-ACHIEV model. A matrix was developed, whereby each candidate indicator was evaluated in relation to the eight criteria, with the objective that a valid and reliable indicator should fulfil as many of these criteria as possible.

Guidelines on the Development of Indicators

A main objective in designing the model was to identify indicators that would have the broadest possible applicability. The extent to which the resultant testing was able to confirm the universality of the model with only very minor alteration evidences the rigour and forward thinking adopted in its initial development.

The identification of appropriate indicators required consideration for the wide spread of tourism activities. The process began with a list of 211 *candidate* indicators (having eliminated parameters which were deemed to be entirely unfeasible or impractical) and to reduce this to a manageable group a number of procedures were followed. This iterative process reduced the list to the 33 indicators demonstrated in Fig. 11.2.

Table 11.1 MEANS criteria for viable indicators

Relevance: The indicator should cover themes and issues which have a significant impact on the development of sustainable tourism;

Availability: Crucial for an indicator is its actual existence, i.e. it must be quantified at regular intervals and the cost of collecting measurements should not outweigh the usefulness of the indicator;

Meaning: A good indicator must be clearly defined and understood without ambiguity by everyone who uses it. The indicator should accurately reflect the concept to be measured, i.e. what is called the validity of construction;

Freshness: The relevant information should be reasonably regularly available and it should be available at the time it is required;

Sensitivity: The quantity in question should be directly responsive to the activity whose performance is being measured, and ideally changes in the quantity should be directly attributable to the activity in question;

Reliability: The same measure undertaken by two different people should produce the same indicator;

Comparability: Ideally an indicator would allow for comparison across a range of different areas, particularly when used for resource and location decisions;

Normativity: Any value given to an indicator should be comparable to a norm, i.e. it should be amenable to setting of benchmarks against which outcomes can be compared.

Source: European Commission, 1999.

As mentioned above, the refining process involved consultations with a broad range of experts including tourism and environmental managers at all levels, planners, enterprise development professionals, heritage and arts professionals, community-based practitioners, tourism industry personnel, scientific experts and expert academics, in addition to consideration of academic and professional literature. The following guidelines for indicator development and best practice were also consulted:

- WTO indicator development guidelines (WTO, 2004)
- UNEP guide: *Making Tourism More Sustainable: A Guide for Policy Makers* (2005)
- The *VICE model* (TMI, 2003)
- *Getting It Right: Monitoring Progress Towards Sustainable Tourism in England* (DCMS, 2001)
- EU MEANS criteria (European Commission, 1999)
- OECD's 'Pressure-State-Response' framework (OECD, 1993) and EEA Driver-Pressure-State-Impact-Response (DPSIR) framework (Smeets & Weterings, 1999)

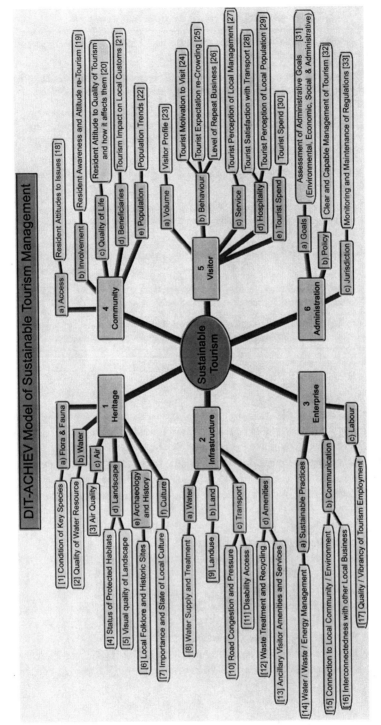

Fig. 11.2 DIT-ACHIEV model of sustainable tourism indicators

Having considered the various frameworks and experts, a number of characteristics of indicators emerged for the development of this model. Cognisance was taken of various indicator types as espoused by Putnam:

- Environmental condition indicators
- Environmental performance indicators (which includes the following two classifications)
- Management performance indicators
- Operational performance indicators (Putnam, 2002)

Indicators were chosen which would assess conditions, trends and performance. In order to make the model as accessible as possible, consideration was given to the assertion by Morrissey, O'Regan, and Moles (2006) that:

> A degree of simplification is a prerequisite . . . to provide information in a form of practical use to decision-makers and understandable to the community (Morrissey et al., 2006: p. 49)

Where possible, the selected indicators also consider: accuracy, bias, age, verifiability and completeness (Putnam, 2002).

Finally the following overarching principles of indicator applicability were established and then taken into account:

- Valuable indicators must consider long-term collection of data.
- Indicators do not have to be specifically tourism related once they can be used to indicate a healthy state of tourism.
- Indicators must assist in indicating data which is useful and consistent.
- Indicators must indicate change over time.
- Indicators must assist in demonstrating movement ('to' or 'from') relative to specified targets (Butler, 2006).

In the implementation of the model, these five principles have been highly useful in illustrating the key characteristics of indicators.

Irish Policy – Setting the Scene

Having established the parameters for indicator development, the following section provides a brief overview of Irish policy regarding both sustainability and sustainable tourism. The purpose of this is to demonstrate that the DIT-ACHIEV model is not developed in a vacuum. It is mindful of international work relating to sustainable tourism development (such as EC, 2006; TSG, 2007; UNEP, 2003), but in addition takes cognisance of the national context, whereby key actors have been working for over a decade and a half to make the Irish tourism industry more sustainable. The Irish Tourist Industry Confederation (ITIC) report on *Tourism and the Environment* in 1986 was the first document that formally recognised the importance of protecting the environment as a product in Irish tourism. It also highlighted the role the environment had to play in creating a sustainable tourism brand for Ireland.

This philosophy is further reflected in the 1995 national response of the Irish Government to the principles established at the 'Earth Summit' held in Rio de Janeiro in 1992 which proposed that each Local Authority revisit its own policies and practices to assess their sustainability. This commitment was further strengthened by the adoption of a *National Sustainable Development Strategy* in 1997, which had the aim of ensuring that economy and society in Ireland develop to their full potential within a well-protected environment (EPA, 2000).

Moving specifically to the sphere of tourism, the *New Horizons for Irish Tourism: An Agenda for Action*, outlines its vision for Irish Tourism as

> a dynamic, innovative, sustainable and highly regarded sector, offering overseas and domestic visitors a positive and memorable experience beyond their expectations (Tourism Policy Review Group, 2003: p. xiv)

According to this report, '[e]nvironmental conservation must become a central element of tourism policy and its implementation'. Four years later, *Tourism and the Environment: Fáilte Ireland's Environmental Action Plan 2007–2009* went a step further by claiming that:

> The future of Irish tourism is inextricably linked to the quality of the environment. Our scenic landscapes, coastline, rivers and lakes, and cultural heritage are the bedrock upon which Irish tourism has been built. The economic viability and competitiveness of the Irish tourism industry can only be sustained if the quality of these resources is maintained. Now, more than ever, Ireland's tourism industry relies on strong and appropriate environmental policies (Fáilte Ireland, 2007: 3).

Specifically addressing sustainability, this action plan goes beyond the environmental focus of many earlier commentaries and states that

> Tourism, when it is well managed and properly located, should be recognised as a positive activity which has potential to benefit the host community, the local environment itself and the visitor alike. Sustainable tourism requires a balance to be struck between the needs of the visitor, the place and the host community (ibid: 13).

The National Development Plan (NDP) 2007–2013 contextualised and acknowledged the importance of Irish tourism as an indigenous growth industry with high employment intensity. In 2006, revenue from the industry had risen to €5.9 billion, paralleled by a 9% increase in overseas visitors. This was the fifth consecutive year of growth in terms of total visitors and earnings, and potential pressure from such rapid expansion was instrumental in motivating the development of the DIT-ACHIEV model. Although tourism numbers and revenue have dipped in 2008 and 2009, it is still a multi-billion euro industry (€6.3 billion in 2008) and while the industry faces difficult years ahead, it is one of Ireland's most important indigenous industries, accounting for 3.1% of GNP annually (Fáilte Ireland, 2009; Tourism Renewal Group, 2009).

Development of New Indicator Model

At a theoretical and conceptual level, Glasson et al.'s Carrying Capacity Web (1995) was highly useful in informing the initial research. However, when it came to examining the Irish context in detail, like many of the other existing models it proved to be restrictive in identifying a robust and all-encompassing set of indicators which would comprehensively consider and satisfy the requirements of local actors and industry. Consequently, a new framework, the *DIT-ACHIEV Model for Sustainable Management of Tourism*, was developed (see Fig. 11.2/Table 11.2) and structured as follows:

- A core comprising six 'fields of interest' which are numbered 1–6 – Heritage, Infrastructure, Enterprise, Community, Visitor and Administration.
- Each of these fields of interest is composed of between 3 and 6 'sub-fields' which are identified by letters (a)–(f).
- As appropriate, the sub-fields are sub-divided into between 1 and 3 indicators.

Thus, there are 33 indicators which are numbered in the diagram using square brackets.

Deliberate Errors?

Two deliberate decisions have been made in the development of the DIT-ACHIEV model, which have been somewhat contentious in certain quarters. First, no attempt has been made to apply weightings to the various indicators. This was an intentional decision so that the model can be applied in any destination. Ultimately it is designed to be employed in a flexible manner, and adapted to the local situation, on a case-by-case basis according to sound environmental, economic and socio-cultural principles. This level of objectivity would be compromised if weightings were predetermined.

Second, a deliberate decision was made to omit the words 'economy', 'society' and 'environment' from the model. This was an attempt to avoid an over-emphasis on these spheres of investigation, which is common in many examinations of tourism impacts. However, economic, social and environmental indicators are implicitly present throughout the model.

Using Indicators

The DIT-ACHIEV model can be utilised to demonstrate different degrees of sustainability, depending on the location/situation in which it is adopted. As suggested earlier, indicators can be used for a number of purposes and primarily in tourism they are used to identify the following:

Table 11.2 DIT-ACHIEV model of sustainable tourism indicators

Fields of interest	Sub-fields	Indicators
(1) Heritage	(a) Flora and fauna	[1] Condition of key species
	(b) Water	[2] Quality of water resource
	(c) Air	[3] Air quality
	(d) Landscape	[4] Status of protected habitats
		[5] Visual quality of landscape
	(e) Archaeology and history	[6] Local folklore and historic sites
	(f) Culture	[7] Importance and state of local culture
(2) Infrastructure	(a) Water	[8] Water supply and treatment
	(b) Land	[9] Landuse
	(c) Transport	[10] Road congestion and pressure
		[11] Disability access
	(d) Amenities	[12] Waste treatment and recycling
		[13] Ancillary visitor amenities and services
(3) Enterprise	(a) Sustainable practices	[14] Water/waste/energy management
	(b) Communications	[15] Connection to local community/environment
		[16] Interconnectedness with other local business
	(c) Labour	[17] Quality/vibrancy of tourism employment
(4) Community	(a) Access	[18] Resident attitudes to issues
	(b) Involvement	[19] Resident awareness and attitude re-tourism
	(c) Quality of life	[20] Resident attitude to quality of tourism and how it affects them
	(d) Beneficiaries	[21] Tourism impact on local custom
	(e) Population	[22] Population trends
(5) Visitor	(a) Volume	[23] Visitor profile
	(b) Behaviour	[24] Tourist motivation to visit
	(c) Service	[25] Tourist expectation re-crowding
	(d) Hospitality	[26] Level of repeat business
	(e) Tourist spend	[27] Tourist perception of local management
		[28] Tourist satisfaction with transport
		[29] Tourist perception of local population
		[30] Tourist spend
(6) Administration	(a) Goals	[31] Assessment of administrative goals (environmental, economic and administrative)
	(b) Policy	[32] Clear and capable management of tourism
	(c) Jurisdiction	[33] Monitoring and maintenance of regulations

- Stresses on an area (e.g. traffic congestion, water shortages, visitor dissatisfaction)
- The impact of tourism (e.g. seasonality, rate of habitat damage, local community employment quality)
- Management effort (e.g. funding of pollution cleanups)
- The effect of management actions (e.g. changed water quality, number of returning tourists)

Table 11.3 Methodology for the formulation of a sustainable tourism strategy

1. Create a multi-interest working group
2. Agree on initial issues to investigate
3. Undertake wide consultation
4. Prepare a situation analysis, including destination performance, needs and opportunities
5. Consult and agree on key issues and priorities
6. Determine strategic objectives
7. Develop an action programme
8. Establish or strengthen instruments to facilitate implementation
9. Implement actions
10. Monitor results

Source: Denman, 2006.

They can also provide an early warning when a policy change or new action may be needed as well as provide a basis for the long-term planning and review of tourism (UNEP/UNWTO, 2005).

An over-arching concern for the project team from the outset of this project has been to ensure that the model would promote a holistic view of tourism impacts. Thus, it was vital to implement a research methodology which would systematically engage with the model in a mature and calculated manner. Denman (2006), in presenting the key themes of the aforementioned *Making Tourism More Sustainable*, proposed a sequence of steps for the formulation of a sustainable tourism strategy. This sequence, which is outlined in Table 11.3, is recommended as a useful tool for this very purpose.

This methodology addresses two key issues. First, it considers establishment of the correct structures through which organisations can work with others towards more sustainable tourism, and thereby develop and drive policies and actions. Second, it highlights that a process must be developed which embraces sustainability and identifies some of the strategic choices that need to be made. Thus, following this format, the DIT-ACHIEV model not only identifies indicators but can be used to actively influence tourism development. Without this latter consideration the model would remain an academic exercise rather than a dynamic and useful tool for sustainable tourism.

Part B: Implementing the Model

Having outlined the development of the DIT-ACHIEV model, the following section presents an overview of its implementation and comments on some of the challenges which the research team have encountered in turning this from theory into a working tool for the promotion of sustainable tourism management. The current phase of research is entitled 'ACHIEVing Sustainable Tourism Management – Putting the DIT-ACHIEV Model into Practice'. The aim is to pilot, test and refine the model and thereby prepare it for implementation by a local community. This current activity will establish whether it has the potential to become a national benchmark tool for the implementation and management of sustainable tourism. From the outset, this is a very personal reflection of the process of implementing the model and focusses on the engagement of actors in this community-focussed work, rather than providing major insight into the workings of the indicators and resultant data.

Selection of Study Area

Once it was developed, it became evident that the model had potential as a tool for benchmarking and evaluating tourism-related changes in administration, community, heritage, infrastructure, enterprise and visitor – particularly at a destination level. As it is acknowledged that all aspects of the model could and should be applied to reach truly sustainable tourism, a case study area/local team would need to have the interest, capacity and capability to engage in these many and diverse areas of investigation.

Rather than randomly select a destination and then attempt to engage local stakeholders, the research team utilised a competitive public tender process, whereby destinations were invited to apply to work with them. Thus, the applicant destinations had to show initiative and desire to partner the DIT in this work.

The shortlisting and eventual selection of a case study area was based on criteria which the applicants had to consider in their respective destination. The selected area/applicant would have to:

- be already considering the future sustainability of the existing tourism product,
- be contained within a clearly defined boundary,
- possess a traditionally strong tourism offering,
- contain defined sites of tourism interest,
- possess a tourism management organisation willing to partner the project team in undertaking the research project,
- have tourism management and organisational capacity,
- show a clear team commitment for a 3-year project,
- display broad community representation and participation,
- demonstrate a knowledge of Fields of Interest and recognition of sustainability issues in general and specific to the location (linked to the DIT-ACHIEV model), and
- exhibit evidence of past achievement(s) in projects of this nature.

A number of destinations were shortlisted and invited to present their case to an expert panel containing members of the project team, international experts, and representatives of the funding agencies.

Selection of Test Site(s)

In considering the metrics for site selection, three fundamental requirements emerged for the potential destination(s): range and extent of physical tourism product; level of stakeholder involvement/engagement; and understanding of challenges to sustainable tourism. While the purpose of the project is to improve the sustainability of a potential destination by quantifiably and qualitatively considering the tourism product, a related objective was to examine whether or not the model could be implemented at a local level. Thus, to a certain extent, the objective was to examine whether sustainable tourism can be led by local organisations and not solely by external 'expert' agencies and forces.

The initial proposal was to test the model in a single destination. However, due to the very high calibre of applicants and the potential benefits of broadening the testing process, it was decided that their level of enthusiasm and capability regarding the project could be capitalised upon. It was therefore decided to expand the parameters and work with two destinations. Using the criteria outlined above, the sites chosen to pilot the model were:

- *Killarney* as a major tourism destination, and
- *Carlingford and Cooley Peninsula* as a minor tourism destination (see Fig. 11.3).

Killarney in the southwest of Ireland is one of the country's major tourism destinations, with a history of visitors frequenting the town and surrounding area since the mid-1700s. This rurally located tourism town has an infrastructure of hotels and other product providers which is second only to Dublin; however, the nearby national park means that awareness of nature and wildlife are always important in planning for tourism. This destination constantly faces challenges of an intensive tourism product in a highly valued and protected environment. In Killarney, project work began with the local actors in January 2009.

Carlingford and the Cooley Peninsula, located north of Dublin, possess a more modest but emerging tourism product. This is a young coastal destination with emphasis on environmentally focussed activity, both water- and land-based. Again, tourism in this location is highly tuned to the environment, and sustainability is foremost in the minds of local actors. Project work in the Carlingford and Cooley Peninsula area began in January 2010 – building on the learnings gained in Killarney.

Implementing the Model with Stakeholders and Actors

Because of the systematic approach employed in the identification of a destination with an appropriately appointed management team, from the outset the project began with an existing multi-interest working group (as per Stage 1 of Denman's Model in Table 11.3). In fact, existing organisations were the drivers and one of the key strengths in both instances.

In each destination the project team established a steering committee which included the team which bid to attract the project; a number of other relevant organisations; a co-ordinator and researcher; academic/technical co-ordinators; and the project funding agencies. Setting up this committee and its regular meeting has been paramount to the success of the project to date. The schematic in Fig. 11.4 visually presents the goals and main activities of the main project partners/steering committee members.

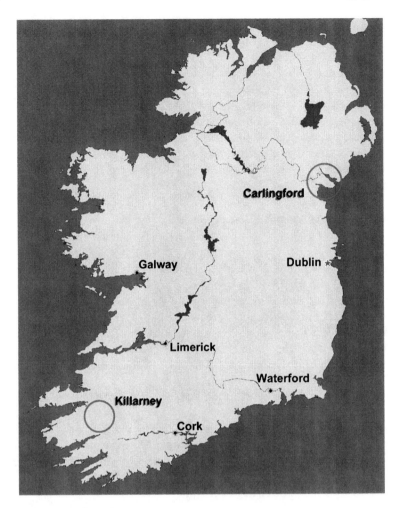

Fig. 11.3 Location of test sites

Goal and Purpose of Each Stakeholder

At the core of this schematic is the DIT-ACHIEV Model of Sustainable Tourism Management, linked to Denman's 10-stage methodology for the Formulation of a Sustainable Tourism Strategy. Around the theoretical core are the all-important stakeholders – in this instance, the local community (involving industry, administration representatives, community activists, etc.), the funders, and the DIT. The model also identifies the key activities for each stakeholder and the key goals they wish to reach in working on the project. These are detailed in the following sections.

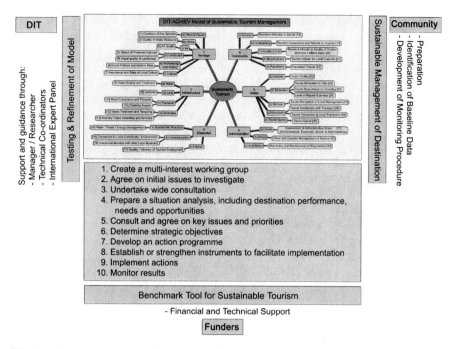

Fig. 11.4   Schematic representation of stakeholders, their involvement and goals

DIT (Dublin Institute of Technology)

The DIT (School of Hospitality Management and Tourism at the Dublin Institute of Technology) is the project leader (see Fig. 11.4). DIT's involvement is both academically and educationally motivated, with a major focus on serving industry needs. As indicated in the diagram, the key goal of the DIT team is – to test and refine the indicator model. Overall, the main DIT engagement centres on providing support and guidance through three main channels:

- Manager/researcher
  The school (via support from the funders) has employed both a full-time research co-ordinator and a part-time project manager. The role of these individuals is to co-ordinate the day-to-day implementation of the model and also to constantly review the processes and activities being undertaken, with a view to streamlining, improving and ultimately developing guidelines for its wider implementation.
- Technical co-ordinators
  Staff/faculty from the school were the primary investigators in the initial development of the model. These individuals are still intimately involved in every stage of the research. Their key roles include activities such as maintaining academic standards, engaging in dissemination of the model and overseeing the activities of the manager and researcher.

- International expert panel

  Over the years, through a number of research projects, and other academic activities, the school has established a network of international experts in the area of sustainable tourism. The panel has included Richard Butler, John Swarbrooke, John Tribe, Frederic Dimanche and Rebecca Hawkins who have all worked on this model. In addition, industry-focussed partners from across Ireland have also been involved. Throughout the various strands of the project, an expert panel has been used as an executive consultation group, and their input has helped to ensure objectivity and an international standard of work.

Project Funders

As stated at the outset, the co-funders (Fig. 11.4) for the project are the Irish Environmental Protection Agency and Fáilte Ireland (the National Tourism Development Authority of the Republic of Ireland). Both of these statutory organisations have very clear objectives within the Republic of Ireland which led to their engagement in this project. The EPA is tasked with achieving balanced and sustainable protection and management of the environment. Through their research and development programme, they generate knowledge and expertise which is relevant in this regard (EPA, 2008b).

In recent years, Fáilte Ireland as part of its remit has been actively advocating a high-quality physical environment for tourism and promoting good environmental practice throughout the sector. In 2005 they established an environment unit which is now centrally located as part of their policy and planning unit (Fáilte Ireland, 2007). Since its inception, this unit has worked with a range of other agencies/bodies such as the EPA, and this partnership approach was behind their cooperative support for the DIT-ACHIEV project. One of the fundamental goals for these two organisations is the development of a tool for benchmarking sustainable tourism in Ireland.

An important observation is that both agencies offer much more than just funding to this process; in fact they have provided technical advice and support throughout. In the various phases of the project work, they have been active participants – steering and guiding the project, while still allowing the researchers/project the freedom to investigate and explore the model in a creative manner.

Community/Local Actors

The group which attracted the researchers to work with Killarney – Team Killarney – was put together by Killarney Chamber of Tourism and Commerce. The team represents the key tourism actors in the area and also includes Killarney National Park, Killarney Town Council, the Trustees of Muckross House and Kerry County Council. This latter organisation/local authority was not initially a member of the consortium, but was included shortly after the project commenced.

The Multi-Interest group in Carlingford is co-ordinated by the Carlingford and Cooley Peninsula Tourism Association, and following the example of Killarney, now includes representatives of a number of local agencies and organisations. The key goal for the community/local groups in each study site is to accomplish sustainable management of tourism in their respective destinations. The main activities which the model suggests groups to undertake include:

- Preparation,
- Identification of baseline data,
- Development of monitoring procedure.

The following sections will provide more information on the activities undertaking in implementing the model, and how the various stakeholders have engaged in the process.

Consultation

Referring to Denman (see Table 11.3), the next two phases of the recommended methodology are:

- agree on initial issues to investigate and
- undertake wide consultation.

Agreeing on initial issues to investigate was somewhat overtaken by the nature of the competitive application process used from the outset of the project. Each of the candidate destinations had to identify initial issues to investigate, and had already mapped these against the DIT-ACHIEV model.

However, extensive public consultation was undertaken in Killarney in order to fulfil two objectives. First, to involve as broad a range of local groups, organisations and interests as possible, and thereby identify a comprehensive breadth of issues, challenges and concerns for tourism. Second, and related to the first, as this was the first time the model was being tested, it was important to investigate its robustness in capturing the key challenges and difficulties in maintaining a sustainable destination. A very intense public consultation process was undertaken, and it is considered that this level of engagement would not be required in other destinations going forward. Over 600 individuals were consulted throughout the extensive range of public meetings and consultations with local agencies, organisations and groups. The process was broadly advertised in local newspapers and radio, and invites and e-mails were also sent seeking submissions from interested parties.

In order to classify and evaluate the range of emerging issues, the many submissions were compiled and mapped against the fields of interest/indicators in the model. The robustness of the model throughout this process was impressive, and while some minor refining was undertaken where it became apparent that the original parameters were too specific, the public consultation process endorsed and supported the extent and range of issues in the model, thereby paving the way for its use as a benchmark tool for sustainable tourism. In Carlingford, the consultation process – informed by Killarney – is currently under way, and a much more refined procedure is being followed. Following its completion, this process will be evaluated and a best practice model will be designed.

Situation Analysis/Identification of Key Issues and Priorities

Having mapped the local issues against the indicator model and established the indicator data that need to be assembled, a local organisation then needs to undertake a range of primary research activities. For this purpose the research team has developed a number of research tools, which are currently being refined and tested. The overarching principle is that common data collection methodologies must be employed to facilitate comparability and testing of data. If these validated tools are used from the outset, key issues and priorities (Phase 5 for Denman's methodology) should emerge organically when results are placed against either the historical benchmarked data in the destination under investigation or data in other similar/comparable locations.

The following diagram illustrates the research tools that have been developed and the areas of the DIT-ACHIEV model which they are used to interrogate. One of the interesting findings to date is that undertaking the various elements in a sequential manner leads to more informed data collection. Thus, the community/resident survey helps inform the visitor survey, and subsequently, information from each of these can assist in focussing the business survey. If all of these data collection methods have been undertaken, an investigation of the local administration will be more informed, and thus the identification of key issues and priorities will be a natural outcome from the various findings.

(A) *Collection/Collation of Pre-existing Quantitative Data*

In selecting indicators, it was important to ensure that data under the categories of heritage and infrastructure are primarily composed of publicly available information. The main difficulty, however, has been first to identify if and where the data are recorded and, second, to gain access to the data. Scientific and other quantitative data can be very straightforward to collect, but experience shows that unless buy-in is received from a multitude of local agencies and organisations, all of the data under this category can be problematic. To collect information in this category, it is important for a local group to forge links with the local authority and key stakeholders such as wildlife rangers, heritage groups and road safety agencies.

(B) *Community/Resident Survey*

In general, the attitudes and opinions of local residents are rarely recorded/documented, and even more so when one focuses on a theme as specific as tourism. To solve this difficulty, a resident survey has been developed. It is proposed that a core set of questions remains constant in all areas where the survey will be undertaken, but if a local organisation/agency wishes to seek responses on other issues, the questionnaire can be adapted/altered to suit. A number of methodologies have been tested (see Fig. 11.5) in undertaking this vital research element of the DIT-ACHIEV model.

In the initial development of the model, a postal survey was undertaken to trial this tool. This proved to be successful, but costly in terms of postal and printing charges. In Killarney, an online community survey was piloted, with the support of all local community organisations. It was thought that this would be useful in a number of ways, not least being the elimination of data-entry costs, but also anonymising the data collection, and theoretically making the whole process more immediate by streamlining its administration. However, despite claims that Ireland is now in the information age, encouraging the local population to complete a web-based survey proved to be very difficult. Multiple techniques were employed, ranging from distributing flyers to houses and businesses, e-mailing various local databases, using Facebook and other social media, adverts in local papers and on radio and even the local library dedicating a computer terminal for people wishing to complete the survey. Despite all of this (and a very attractive family prize) it took considerable effort to achieve a statistically representative completion rate. To be inclusive, printed versions of the survey were distributed from the outset, and these were important in achieving the target results. In Carlingford, a two-pronged approach has been taken: local individuals have been trained as interviewers and also a group of postgraduate students undertook a number of days of surveying. The purpose of

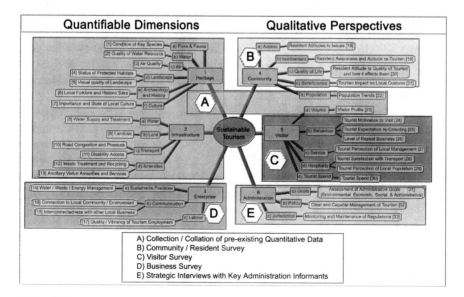

Fig. 11.5 DIT-ACHIEV model – research methodologies

this dual methodology are both pragmatic and research oriented in nature. The student participation will assist the local population to achieve their target number of completed surveys, but also their results will act as a means of testing and validating the findings of the local volunteers. In discussion at steering committee level, it was posited that local surveyors could unintentionally influence the findings and thus, the research team will compare the two datasets to investigate local bias. If this analysis validates the findings of the locally administered surveys, it is suggested that this methodology will be adopted for the model going forward.

(C) *Visitor Survey*

A detailed visitor survey is an essential element of data collection for the model (see Fig. 11.5). Again, detailed tourism information of a local nature is difficult if not impossible to source, and to attain a detailed understanding of the visitor, survey work needs to be undertaken. As with the resident survey, a questionnaire has been designed, and through its various iterations, a common set of questions has become core. These can be altered slightly, particularly if deemed necessary in light of findings from the resident survey.

In the initial project, a researcher was employed to undertake the visitor surveys. Going forward, an expense of this nature would be completely impractical for any destination. Thus, a re-think was required. The method which has been piloted in Killarney has been very successful thus far. A broad range of tourism businesses (various accommodation types and attractions, etc.) volunteered to complete a small number of questionnaires with their customers each week. Thus, with 15 businesses taking part and each one undertaking two surveys per week, 1,560 surveys could be completed in the year. Because some businesses close during the winter, or at times staff can be too busy to spend time on the survey, a certain latitude has to be provided. Once a commitment is received, this methodology can produce very rich results indeed, with a minimum of effort from each of those involved. In Killarney, it has been suggested that after this project has finished, the local Chamber of Tourism and Commerce will continue to sponsor and administer this particular research element.

To facilitate future development of this tool, a number of methodologies for data analysis are being tested. The survey could migrate to a user-completed online version, or it could remain in its current format with the interviewers entering their surveys online each week. Either of these methodologies would facilitate an automated production and analysis of results at little or no cost to the organisation or group co-ordinating the project.

(D) *Business Survey*

The Business Survey (see Fig. 11.5) is one of the most straightforward elements of the model, and because most businesses are now technologically aware, an online survey is proving to be far more successful with this group of participants than with the general public. In addition, it is in the interest of many of the participants to respond to the survey, as the findings should provide results that will inform and benefit their businesses in the future.

(E) *Strategic Interviews with Key Administration Informants*

From development of the model and previous research involving the local administration, the research team advocate the use of strategic conversations with key informants for investigating the administration field of interest (see Fig. 11.5). In order for this to be fruitful, the findings of the previous research tools will feed into this methodology. Strategic conversations are highly useful in extracting meaning from key informants regarding the operation and management of local issues, challenges and goals (Coccossis & Mexa, 2004; Van der Heijden, 1996). Thus, benchmarking of indicators in the areas of goals, policy and jurisdiction will be efficiently established through this approach.

Community Management of Model

As stated at the outset, this is an ongoing project, which has evolved from a very academic exercise towards becoming a hands-on, toolkit-style methodology for the sustainable management of tourism at a destination level. While work has not yet progressed sufficiently to provide detailed insights into how the latter stages of Denman's methodology will shape the methodology, the model is already providing practical direction and guidance for the management of tourism in two Irish destinations. It is suggested that when one considers stages 6–10 in Denman's methodology, the main challenge going forward will be developing sufficient capacity at a destination level, to facilitate an analytical and reflective approach to determining strategic objectives. Once this has been achieved, the remaining steps in the process are pragmatic in nature and evolve organically. Thus, once a strategic approach is attained, and the indicators have been populated with valid and sound data, implementing the model should be achievable.

These final phases of the methodology are well within the capacity of the organisations that have engaged with the research team thus far. Therefore, the final challenge will be to develop a methodology for determining strategic objectives in a holistic and all-encompassing manner, which will be repeatable in all destinations – even ones where the management team may not be as strong as the partners who have engaged with the model to date. Initial steps in this regard have been very promising, and it would appear that with detailed and practical guidelines/structures the indicators within the model can evolve and develop into strategic objectives to lead a destination to a sustainable future.

Local communities who employ this model will be particularly eager to reach the final two phases of Denman's methodology – the implementation of actions and monitoring results. Once they have reached this stage of the process, they will be able to see the fruits of their labours and begin a process of feeding results back into the model. It is only when this point has been reached that the true value of the model will be realised, and ongoing evaluation and reflection will be attainable.

Conclusion

The main challenge for the research team leading this project is to work towards stepping away from leading/controlling the work and allowing community management of the DIT-ACHIEV model. In the long term, it is proposed that the model will be rolled out in destinations across the country and become a national benchmark tool for sustainable tourism management. To make this a reality, structures/methodologies/support mechanisms are being developed whereby local groups can utilise the model and its related tools with a minimum of external support. Communities must be empowered to utilise the model to the full, and thus, writing themselves out of the methodology in a real and viable manner is an interesting and ongoing challenge for the project team. This can only be achieved, however, if local community actors and stakeholders continue to engage in the process of making the management of their own tourism destinations as sustainable as possible.

References

Butler, R. (2006) One Day Indicator Refining Workshop with Professor Butler [unpublished], cited in Flanagan, S., Griffin, K., O'Halloran, E., Phelan, J., Roe, P., Kennedy-Burke, E., Tottle, A. & Kelly, R. (2007) *Sustainable Tourism Development: Towards the Mitigation of Tourism Destination Impacts*, Environmental Protection Agency, Wexford Ireland.

Coccossis, H., & Mexa, A. (2004). *The challenge of tourism carrying capacity assessment: Theory and practice*. Aldershot: Ashgate.

DCMS Department for Culture, Media and Sport. (2001). National sustainable tourism indicators: Getting it right: Monitoring progress towards sustainable tourism in England. London: Crown copyright.

Denman, P. (2006). Tourism and Sustainability: Objectives, policies and tools for sustainable tourism, Paper Presented to UNWTO seminar on tourism sustainability and local agenda 21 in tourism destinations, Jeddah, Saudi Arabia, 18 and 19 February 2006.

Environmental Protection Agency (2000). *Ireland's environment: A millennium report*. Wexford: EPA.

Environmental Protection Agency. (2004). *Ireland's environment 2004*. Wexford: EPA.

Environmental Protection Agency. (2008a). *Ireland's environment 2008*. Wexford: EPA.

Environmental Protection Agency. (2008b). *The environmental protection agency who we are – What we do*. Wexford: EPA.

European Commission. (1999). Evaluating socio-economic programmes – Vol. 2, Selection and use of indicators for monitoring and evaluation (part 2 of 6 volume collection). Luxembourg: Office for Official Publications of the European Communities.

European Commission. (2006). Review of the EU sustainable development strategy (EU SDS) – Renewed strategy. Brussels: Council of the European Union.

Fáilte Ireland. (2007). *Tourism and the environment: Fáilte Ireland's environmental action plan 2007–2009*. Dublin: Fáilte Ireland.

Fáilte Ireland. (2009). Tourism facts 2008. Dublin: Fáilte Ireland.

Flanagan, S., Griffin, K., O'Halloran, E., Phelan, J., Roe, P., Kennedy-Burke, E., et al. (2007). *Sustainable tourism development: Towards the mitigation of tourism destination impacts*. Wexford: Environmental Protection Agency.

Glasson, J., Godfrey, K., & Goodey, B. (1995). Towards visitor impact management: Visitor impacts, carrying capacity and management responses in Europe's historic towns and cities. Aldershot: Ashgate Publishing Ltd.

Government of Ireland. (1997). *Sustainable development: A strategy for Ireland*. Dublin: Stationery Office.

Government of Ireland. (2007). National development plan 2007–2013. Dublin: Stationery Office.

ITIC (Irish Tourism Industry Confederation). (1986). *Tourism and the environment*. Dublin: ITIC.

Kelly, R., Sirr, L., & Ratcliffe, J. (2004). The future of sustainable development: A European perspective. In Didsbury, H. (Ed.), *Thinking creatively in turbulent times*. Maryland: World Future Society.

Morrissey, J., O'Regan, B., & Moles, R. (2006), Development of indicators and indices for the evaluation of the sustainability of Irish settlements and regional settlement patterns, In: Proceedings of ENVIRON 2006, University College Dublin, January 2006.

OECD. (1993). *OECD core set of indicators for environmental performance reviews* (OECD Environment Monographs No. 83). Paris: OECD.

Putnam, D. (2002) ISO 14031: Environmental Performance Evaluation Draft Paper submitted to Confederation of Indian Industry for publication in their Journal. September 2002.

Smeets, E., & Weterings, R. (1999). *Environmental indicators: Typology and overview*. Technical report No 25. Copenhagen: European Environment Agency.

TMI (Tourism Management Institute). (2003). Destination management handbook. London: TMI & European Travel Commission.

TSG (Tourism Sustainability Group). (2007) Action for More Sustainable European Tourism, European Commission.

Tourism Policy Review Group. (2003). New Horizons for Irish Tourism: An agenda for action, Report to the Minister for Arts, Sports and Tourism, Dublin.

Tourism Renewal Group. (2009). Report of the tourism renewal group: Survival, recovery and growth – A strategy for renewing Irish tourism, 2009–2013. Dublin: Tourism Renewal Group.

UNEP (United Nations Environment Programme). (2003). Tourism and Local Agenda 21. The Role of Local Authorities in Sustainable Tourism.

United Nations Environment Programme/United Nations World Tourism Organisation (UNEP/UNWTO). (2005). Making tourism more sustainable: A guide for policy makers. Madrid: UNEP Division of Technology, Industry and Economics.

Van der Heijden, K. (1996). *Scenarios: The art of strategic conversations*. Chichester: Wiley.

WTO/UNWTO (United Nations World Tourism Organisation). (2004). Indicators of sustainable development for tourism destinations: A guidebook. Madrid: WTO.

Williams, P., & Gill, A. (1993). Tourism carrying capacity management issues. In Theobold, W. (Ed.), *Global tourism: The next decade*. Oxford: Butterworth–Heinemann.

Photos of Killarney

Photos of Carlingford

Launch of project in Killarney – with local partners

Index

M. Budruk, R. Phillips (eds.), *Quality-of-Life Community Indicators for Parks,
Recreation and Tourism Management,* Social Indicators Research Series 43,
DOI 10.1007/978-90-481-9861-0, © Springer Science+Business Media B.V. 2011

CPSIA information can be obtained at www.ICGtesting.com
Printed in the USA
238964LV00004B/36/P